CHINESE C4I/EW
VOL. 1

WENDELL MINNICK

EDITOR

Copyright © 2022

WENDELL MINNICK, EDITOR

All rights reserved.

ISBN: 9798401716521

DEDICATION

To all the American Old Crows out there!

CONTENTS

CIDEX Background	1
YITU	2
Poly Technologies/Poly Defense	41
China Electronics Corp. (CEC)	51
Luoyang Institute	106
INDEX	121
Other Books	158

ABOUT THE EDITOR

WENDELL MINNICK, BS, MA

Show Coverage:

o Aero India – 2007, 2009, 2011
o Association of United States Army (AUSA) 2006 Annual Meeting – Washington, DC
o China International Aviation and Aerospace Exhibition (Zhuhai Airshow) - 2006, 2010, 2012, 2014, 2016
o Defence and Security 2017 (Bangkok) - 2017
o Defence Services Asia (DSA-Kuala Lumpur) - 2008, 2014, 2016, 2018
o Defence Technology Asia International Conference and Exhibition (Singapore) - 2007
o Dubai Airshow - 2007, 2009
o Global Security Asia (Singapore) - 2007
o International Defence Exhibition (IDEX – Abu Dhabi) – 2007
o International Maritime Defense Exhibition (IMDEX – Singapore) – 2007, 2009, 2011, 2013, 2015, 2017
o Langkawi International Maritime and Aerospace Exhibition (Malaysia) - 2017
o Singapore Air Show – 2006, 2008, 2010, 2012, 2014, 2016, 2018
o Seoul Air Show – 2009, 2017
o Shangri-la Dialogue (Singapore) – 2008, 2009, 2010, 2011, 2012, 2013, 2015, 2016, 2017, 2018
o Taipei Aerospace and Defense Technology Exhibition – 2001, 2003, 2005, 2007, 2009, 2011, 2013, 2015, 2017
o U.S.-Taiwan Defense Industry Conference – 2006 (Speaker)

PUBLICATIONS:

o *Afghanistan Forum*
o *Army Magazine*
o *Asian Profile*
o *Asian Thought and Society*
o *Asia Times*
o *BBC*
o *C4ISR Journal*
o *Chicago South Asia Newsletter*
o *Defense News*
o *European Security and Defence Magazine*
o *Far Eastern Economic Review*
o *International Peacekeeping*
o *Jane's Airport Review*
o *Jane's Defence Upgrades*
o *Jane's Defence Weekly*
o *Jane's Intelligence Review*
o *Jane's Missiles and Rockets*
o *Jane's Navy International*
o *Jane's Sentinel Country Risk Assessments*
o *Japanese Journal of Religious Studies*
o *Journal of Asian History*
o *Journal of Chinese Religions*
o *Journal of Oriental Studies*

- *Journal of Political and Military Sociology*
- *Journal of Security Administration*
- *Journal of the American Academy of Religion*
- *Military Intelligence Professional Bulletin*
- *Military Review*
- *New Canadian Review*
- *New World Outlook*
- *Pacific Affairs*
- *Powerlifting USA*
- *South Asia In Review*
- *Taipei Times*
- *Topics (Taiwan – AMCHAM)*
- *The Writer*
- *Towson State Journal of International Affairs*
- *Writer's Connection*

Book

Spies and Provocateurs: A Worldwide Encyclopedia of Persons Conducting Espionage and Covert Action, 1946-1991 (North Carolina: McFarland, 1992). The book was well received in the U.S. intelligence community, including positive book reviews in *Cryptolog, Cryptologia, Military Intelligence Professional Bulletin, Periscope* (AFIO) and *The Surveillant*. The book was also profiled in the 1995 release of the *Whole Spy Catalog: A Resource Encyclopedia for Researchers*.
Editorial projects include the following reports published under my name:

China Market Outlook for Civil Aircraft, 2014-2033 (2016)
Chinese Aircraft Engines (2016)
Chinese Air-Launched Weapons & Surveillance, Reconnaissance and Targeting Pods (2019)
Chinese Anti-Ship Cruise Missiles (2019)
Chinese Fighter Aircraft (2016)
Chinese Fixed-Wing Unmanned Aerial Vehicles (2016)
Chinese Helicopters (2016)
Chinese Radars (2017)
Chinese Rocket Systems (2016)
Chinese Rotary/VTOL Unmanned Aerial Vehicles (2016)
Chinese Seaplanes, Amphibious Aircraft and Aerostats/Airships (2016)
Chinese Space Vehicles and Programs (2016)
Chinese Submarines and Underwater Systems (2019)
Chinese Tanks and Mobile Artillery (2018)
Directory of Foreign Aviation Companies in China (2014)
I Was a CIA Agent in India: Analysis (2015)
List of Foreign Companies and Identities of Taiwan Local Agents (2019)
More Chinese Fixed-Wing UAVs (2019)
More Chinese Rotary & VTOL UAVs (2019)
Taiwan CyberWarfare (2018)
Taiwan Space Vehicles (2018)
The Chinese People's Liberation Army: Analysis of a Cold War Classic (1950/2015)
Unicorn: Anatomy of a North Korean Front: Casinos, Immigration, Trade Sanctions and Violations (2019)

NOTE TO READER

This book includes product brochures obtained directly from manufacturers at defense trade shows and air show exhibitions between 2000-2018 covering the Middle East and Asia, including the Zhuhai Airshow. The quality of these jpgs was normally 300dpi, as Print On Demand (POD) has restrictions on quantity. Therefore, some of the photographs might appear poor quality. The important thing is the data is genuine, though raw and unvarnished.

RADAR ISSUES

There are some radars listed that did not appear in my earlier Chinese Radars book. If the radar you are looking for does not appear in C4I/EW please consult my 2018 *Chinese Radars* (2018) book available on Amazon for $9.99.
The same is true of optical-pods for aircraft, both manned and unmanned: see for Chinese unmanned aircraft books: *Chinese Rotary/VTOL Unmanned Aerial Vehicles* (2016), *More Chinese Rotary & VTOL UAVs* (2019), *Chinese Fixed-Wing Unmanned Aerial Vehicles* (2016), *More Chinese Fixed Wing UAVs* (2019), and *Chinese Air-Launched Weapons & Surveillance, Reconnaissance and Targeting Pods* (2019).

BACKGROUND ON CIDEX

Summation of *Defense News* Articles

Defense News, 03/19/2008, Beijing Prepares for CIDEX 2008, Wendell Minnick.

TAIPEI - Beijing is preparing for its 6th China International Defence Electronics Exhibit (CIDEX), which will feature the second China Defence Electronics Application Forum. Scheduled for April 2-5 at the Beijing Exhibition Center, CIDEX is organized by the state-owned China National Electronics Import and Export Corp. and managed by China Electronics International Exhibition & Ads. The China Defence Electronics Application Forum will include sessions on remote sensing, radar, guidance techniques, electronic information, electronic intelligence, electronic surveillance measures and communication encryption technologies. Held every two years since 1998, the 2006 CIDEX saw more than 250 exhibitors from China, Czech Republic, Germany, Japan, South Africa, South Korea, Switzerland, Ukraine and the United Kingdom, and more than 20,000 visitors from 28 different countries. Despite the large turnout, most exhibitors were Chinese. The exhibit covers a variety of areas: communication systems, electronic warfare systems, aviation electronics, air defense and control systems, nuclear electronic equipment, laser equipment and optic-electronic devices, anti-terrorism technology, fire control systems, and emulation and training. "Since the informatization is one of the pivotal tasks for the process of China's military build-up, China's investment into the relevant equipment against the background of informatization has been increasing year by year," a CIDEX news release said. "In this case, overseas defense electronics manufacturers and distributors are seeking to find proper channels into the Chinese market, increase their presence and enhance the cooperation with relevant governmental department and local companies. To this end, CIDEX serves as a good platform for their market activities."

Defense News, 04/23/2010, China Prepares for CIDEX 2010, Wendell Minnick.

TAIPEI - The China International Defence Electronics Exhibition (CIDEX 2010) is set for May 12-14 at the Beijing Exhibition Center. Organized by the China National Electronics Import and Export Corp. (CEIEC), CETC International Co., and Beijing Xinlong Electronics Technology Co., CIDEX is the "most professional and authoritative defense electronics exhibition in China, covering both military and civilian applications," show organizers said. Sponsored by the General Equipment Headquarters of the People's Liberation Army, Ministry of Industry and Information Technology, and China Electronics Corp., CIDEX 2010 is expected to break previous records in attendance with more than 300 exhibitors from 13 countries and 20,000 visitors expected this year. "Promoting informationization of China's army will be the core task of the construction of Chinese army in the next five years," show officials said. Therefore, investments in equipment for China's military have been rising steadily each year. "In this context, foreign manufactures and companies specialized in defense electronics are planning to enter the Chinese market and enhance ... influence in China through certain channels."

YITU TECHNOLOGY/POLY TECHNOLOGIES

While attending the Defense Services Asia Exhibition & Conference (DSA) 2018, Kuala Lumpur, Malaysia, 16 Apr 2018 - 19 Apr 2018, the editor of this book (Wendell Minnick) visited two Chinese security/defense industry booths. The below is a summation of the article written about the show for a defense publication. After the below summation are brochures acquired at separate booths for Yitu Technology and Poly Technologies, respectively. Both systems are identical, according to a U.S. intelligence community source attending the DSA who spoke to the author.

Beijing-based Yitu Technology showcased its artificial intelligence (AI) solutions for the first time at that year's DSA. The Yitu booth demonstrated its facial recognition software and hardware products to attendees.

The AI's algorithm is capable of identifying over 1.8 billion faces and has been deployed at events such as the G20 Summit, BRICS Summit and Boao Forum for Asia.

The company has made an impression on the international AI community, including the U.S. government.

In 2017, Yitu placed first in four of six categories of the face recognition vendor test, organized by the National Institute of Standards and Technology (NIST). Also, in 2017, Yitu placed first against international competitors at the Face Recognition Prize Challenge (FRPC).

Note: NIST is under the US Department of Commerce, and FRPC was hosted by the Intelligence Advanced Research Projects Activity (IARPA) under the US Office of the Director of National Intelligence.

After a tour of the booth, a company representative gave a demonstration to the author of this article. After scanning, the system had already recorded his movements using a variety of cameras stationed around the booth, including an eyeglass camera system dubbed Smart Glasses. The system can also integrate surveillance from UAVs and body-worn cameras to help identify 'persons of interest'.

The company displayed an English-language 8 February article by Kuala Lumpur-based media outlet, *The Star Online*, reporting that China's police were now 'sporting' the Smart Glasses to catch criminals. The article added there were concerns being expressed by international human rights groups, though it was unclear if Yitu was aware of what the article was suggesting to visitors at the booth.

Yitu also displayed a portable facial recognition system that folded up into an easy to carry 10kg 'briefcase'. The company did not identify the system by name, but a US defense industry analyst indicated it was also being marketed by the Beijing-based defense company, Poly Technologies, as the FocusBrain system.

Details were hard to acquire from company representatives at the Poly booth, but a brochure provided some specifications. The system is a 'combat type' that provides real-time video for 'end-to-end public security solutions for large, open and crowded events'.

The system is remarkable: milliseconds alarm delay to all connected end devices for a flow rate of 100 people per minute, supports up to 10 million photo database storage, 500,000 'blacklist' database, connects up to 30 FocusBrain systems for joint operations under a built-in Wi-Fi for movable operations and connects up to four security cameras through 'power over ethernet' for live video stream processing. It has a 22-core Intel Xeon Processor, DDR4 128G RAM and SSD 2T hard drive.

According to a Yitu brochure, the company was founded in 2012 and has since received $55 million in 'Series C' funding from Hillhouse Capital Group, Yunfeng Capital, Banyan Capital, US-based Sequoia Capital and the Zhenfund.

In 2017, the company established a strategic cooperation with Microsoft to integrate Yitu's AI solutions into Microsoft's AZURE Cloud Computing Platform for 'Smart City' applications. This year the company launched its first international office in Singapore to better serve the Southeast Asia region. This will include Hong Kong, Macau and Oceania regions, for the time being.

Yitu was one of the many cybersecurity outfits exhibiting at the first National Security Asia (NATSEC Asia) exhibition, which was riding on this week's DSA show. NATSEC Asia featured a cyber defense and security zone, along with a conference on cyber defense.

YITU Booth. FocusBrain. 2018 Defense Services Asia. Author Photograph (1/7).

YITU Booth. 2018 Defense Services Asia. Author Photograph (2/7).

YITU Booth. 2018 Defense Services Asia. Author Photograph (3/7).

YITU Booth. 2018 Defense Services Asia. Author Photograph (4/7).
Note: Author tagged by system.

**YITU Booth. 2018 Defense Services Asia. Author Photograph (5/7).
Note: Author tagged by system three times.**

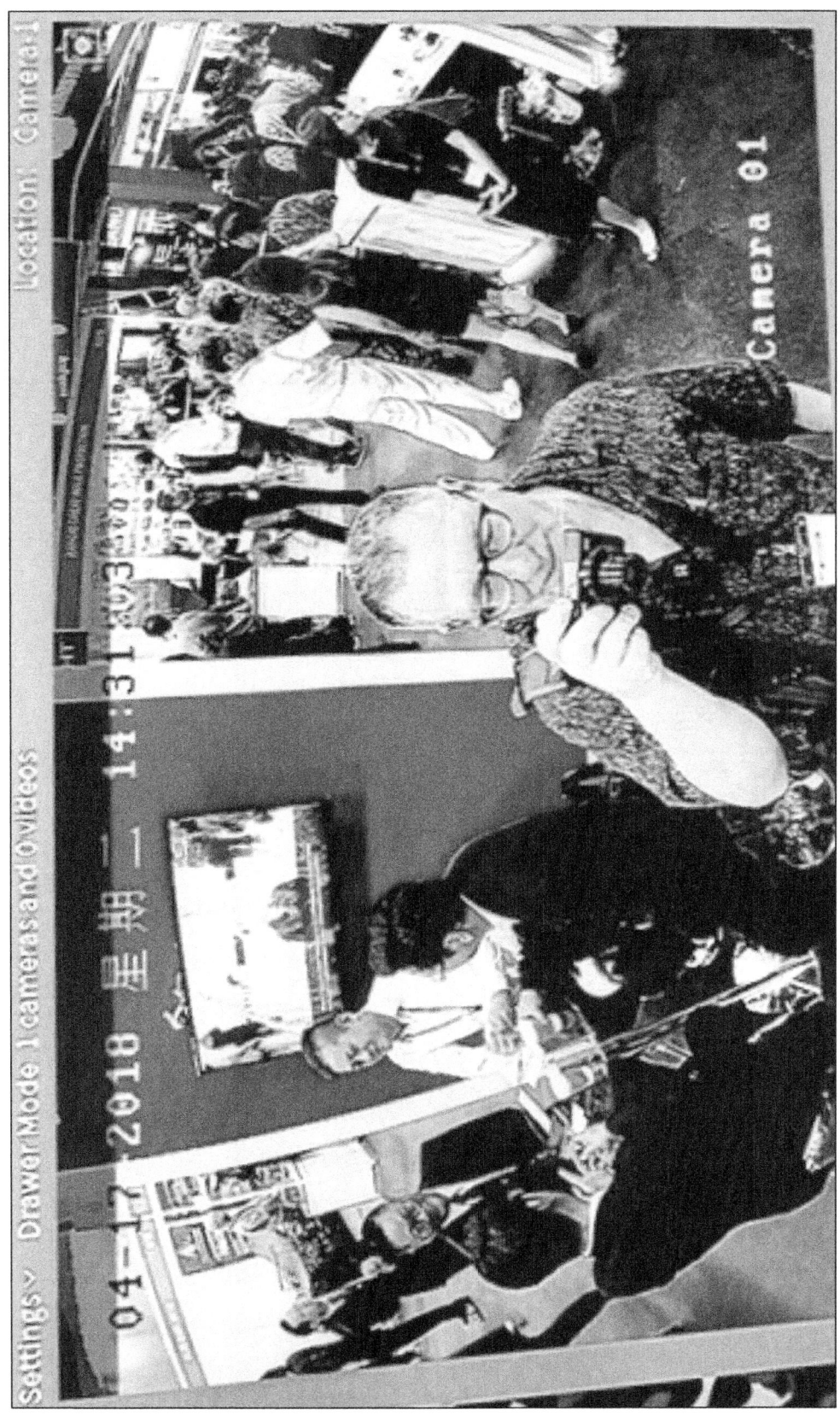

YITU Booth. 2018 Defense Services Asia. Author Photograph (6/7).
Note: Author tagged by system.

YITU Booth. 2018 Defense Services Asia. Author Photograph (7/7).
Note: Author tagged by system three times.

Poly Technology Booth. 2018 Defense Services Asia. Author Photograph (1/4).

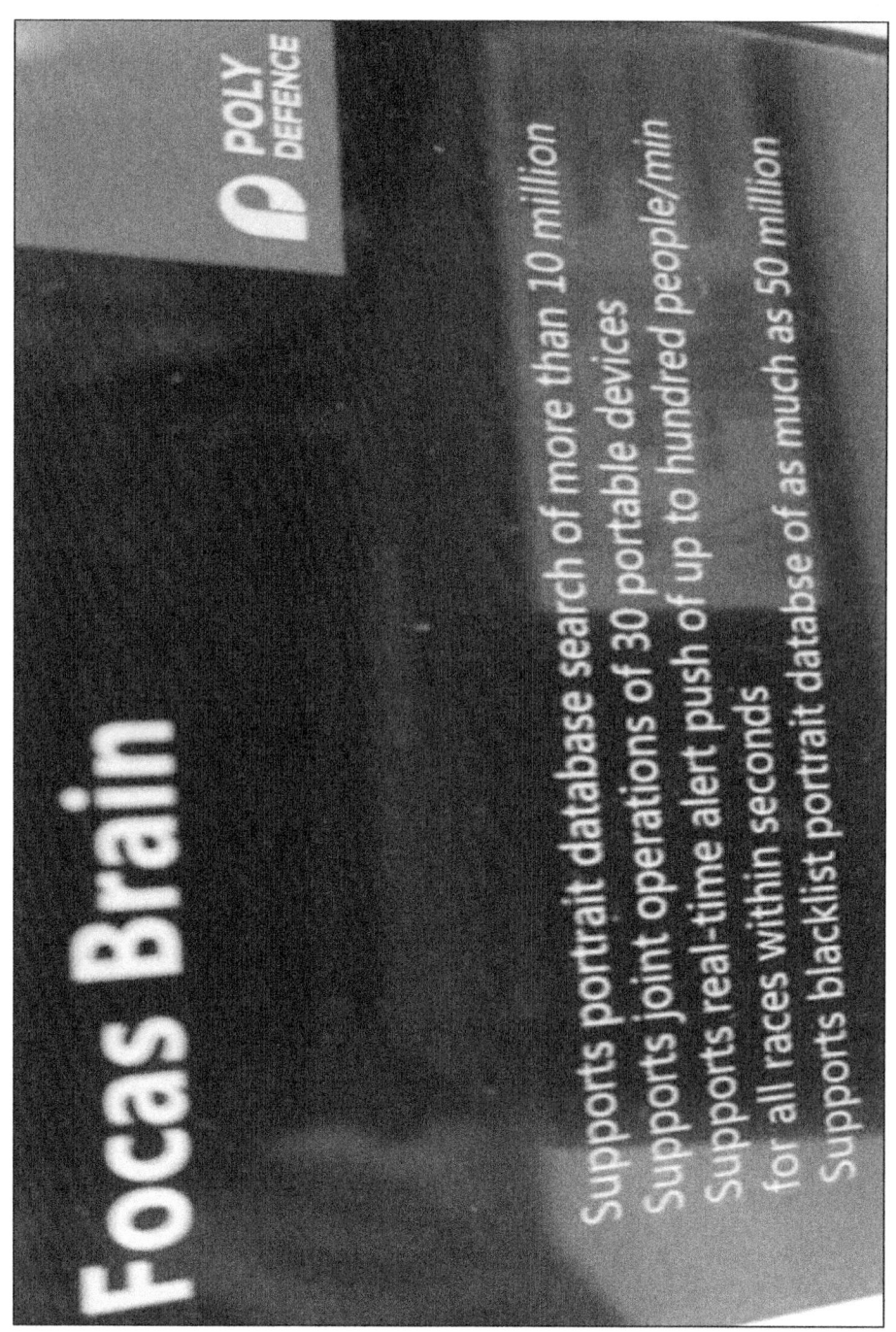

Poly Technology Booth. FocusBrain. 2018 Defense Services Asia. Author Photograph (2/4).

Poly Technology Booth. FocusBrain. 2018 Defense Services Asia. Author Photograph (3/4).

Poly Technology Booth. 2018 Defense Services Asia. Author Photograph (4/4).

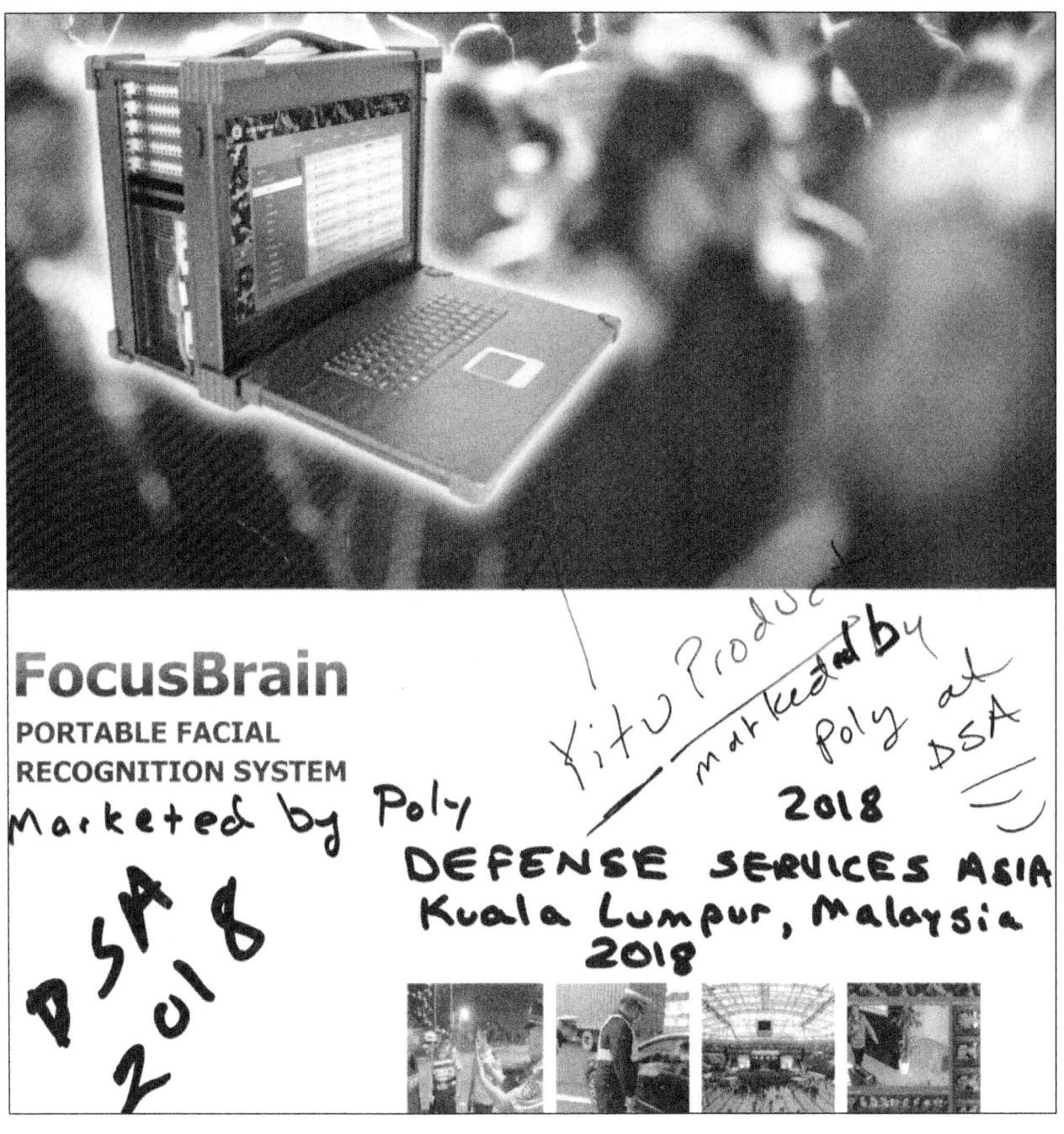

Focus Brain Brochure. Poly Technologies. 2018 Defense Services Asia. (1/4).

GENERAL DESCRIPTION

Configured with the most robust hardware and the world's leading algorithm, FocusBrain is a combat-type, portable facial recognition machine, providing end-to-end public security solutions for large, open and crowded events surveillance scenarios.

FEATURES

- All race facial recognition.
- Supports crowded, populated open space event monitoring.
- Real-time video stream processing.
- End devices real-time alarm push.
- Supports face search and verification.

FUNCTIONS

- Milliseconds alarm delay to ALL connected end-devices for a flow rate of 100 people/minute.
- Supports up to 10 million photo database storage; 500,000 blacklist database for real-time surveillance purposes.
- Connects up to 30 FocusBrain for joint operation under build-in wifi for movable operations and large area coverage.
- Connects up to 4 security cameras through POE for live video stream processing.

SPECIFICATION

CPU	22 core Intel Xeon Processor
RAM	DDR4 128G
Hard Drive	SSD 2T
Screen	17.3 inch FHD 1920*1080
Dimensions	418mm*186mm*352mm
Weight	10kg
WIFI Board	2.4G/5G Double frequency wireless network card
Working Environment	0℃- 50℃

Add: 27/F, New Poly Plaza, No.1 Chaoyangmen Beidajie, Dongcheng District, Beijing, China
Post Code: 100010
Tel: (8610) 6408 2288   Fax: (8610) 6408 2988
www.poly.com.cn   E-mail: poly@polyinc.com

Focus Brain Brochure. Poly Technologies. 2018 Defense Services Asia. (2/4).

GENERAL DESCRIPTION

Configured with the most robust hardware and the world's leading algorithm, FocusBrain is a combat-type, portable facial recognition machine, providing end-to-end public security solutions for large, open and crowded events surveillance scenarios.

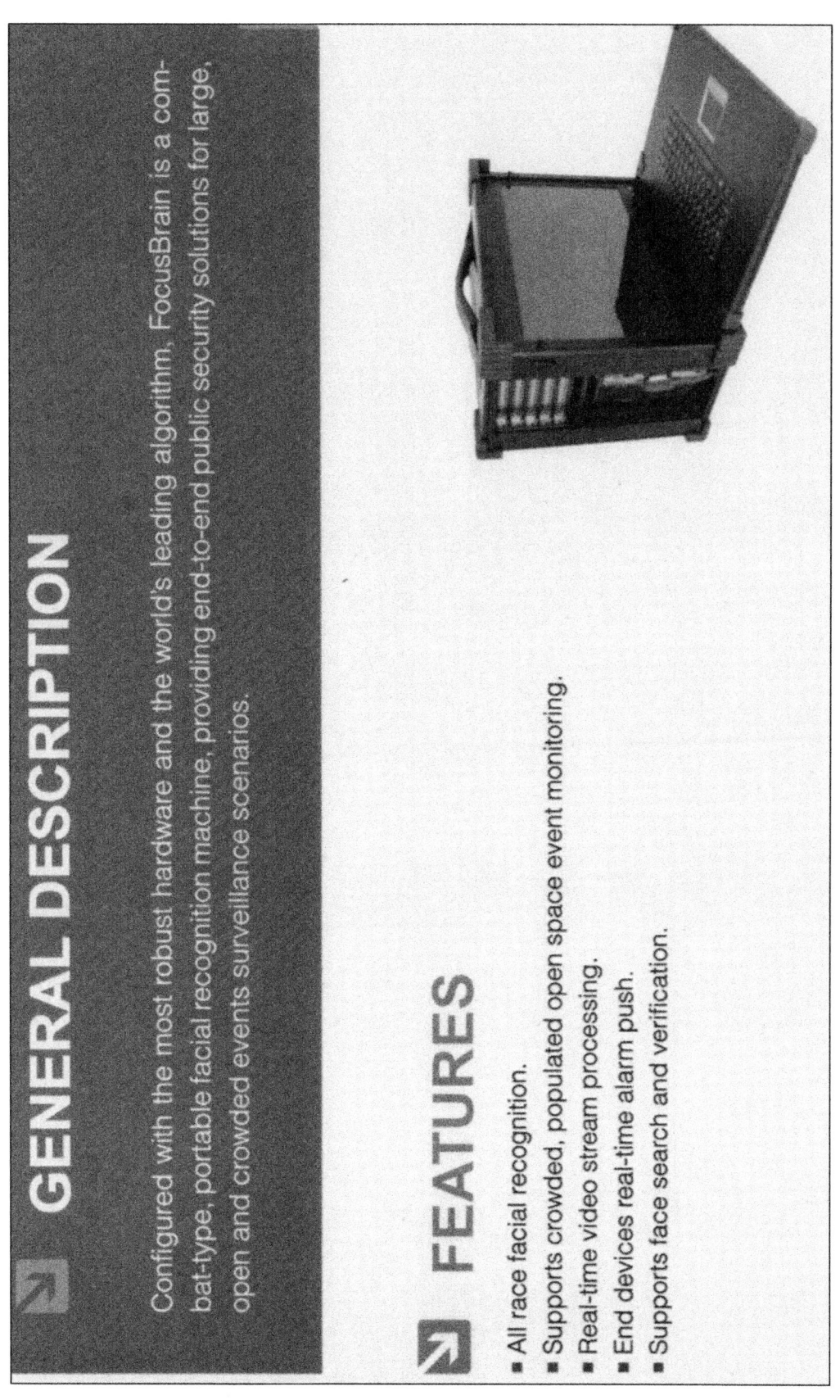

FEATURES

- All race facial recognition.
- Supports crowded, populated open space event monitoring.
- Real-time video stream processing.
- End devices real-time alarm push.
- Supports face search and verification.

Focus Brain Brochure. Poly Technologies. 2018 Defense Services Asia. (3/4).

FUNCTIONS

- Milliseconds alarm delay to ALL connected end-devices for a flow rate of 100 people/minute.
- Supports up to 10 million photo database storage; 500,000 blacklist database for real-time surveillance purposes.
- Connects up to 30 FocusBrain for joint operation under build-in wifi for movable operations and large area coverage.
- Connects up to 4 security cameras through POE for live video stream processing.

SPECIFICATION

CPU	22 core Intel Xeon Processor
RAM	DDR4 128G
Hard Drive	SSD 2T
Screen	17.3 inch FHD 1920*1080
Dimensions	418mm*186mm*352mm
Weight	10kg
WIFI Board	2.4G/5G Double frequency wireless network card
Working Environment	0°C- 50°C

POLY TECHNOLOGIES INC.

Add: 27/F, New Poly Plaza, No.1 Chaoyangmen Beidajie, Dongcheng District, Beijing, China
Post Code: 100010
Tel: (8610) 6408 2288   Fax: (8610) 6408 2988
www.poly.com.cn   E-mail: poly@polyinc.com

Focus Brain Brochure. Poly Technologies. 2018 Defense Services Asia. (4/4).

YITU Brochure. 2018 Defense Services Asia. (1/22).

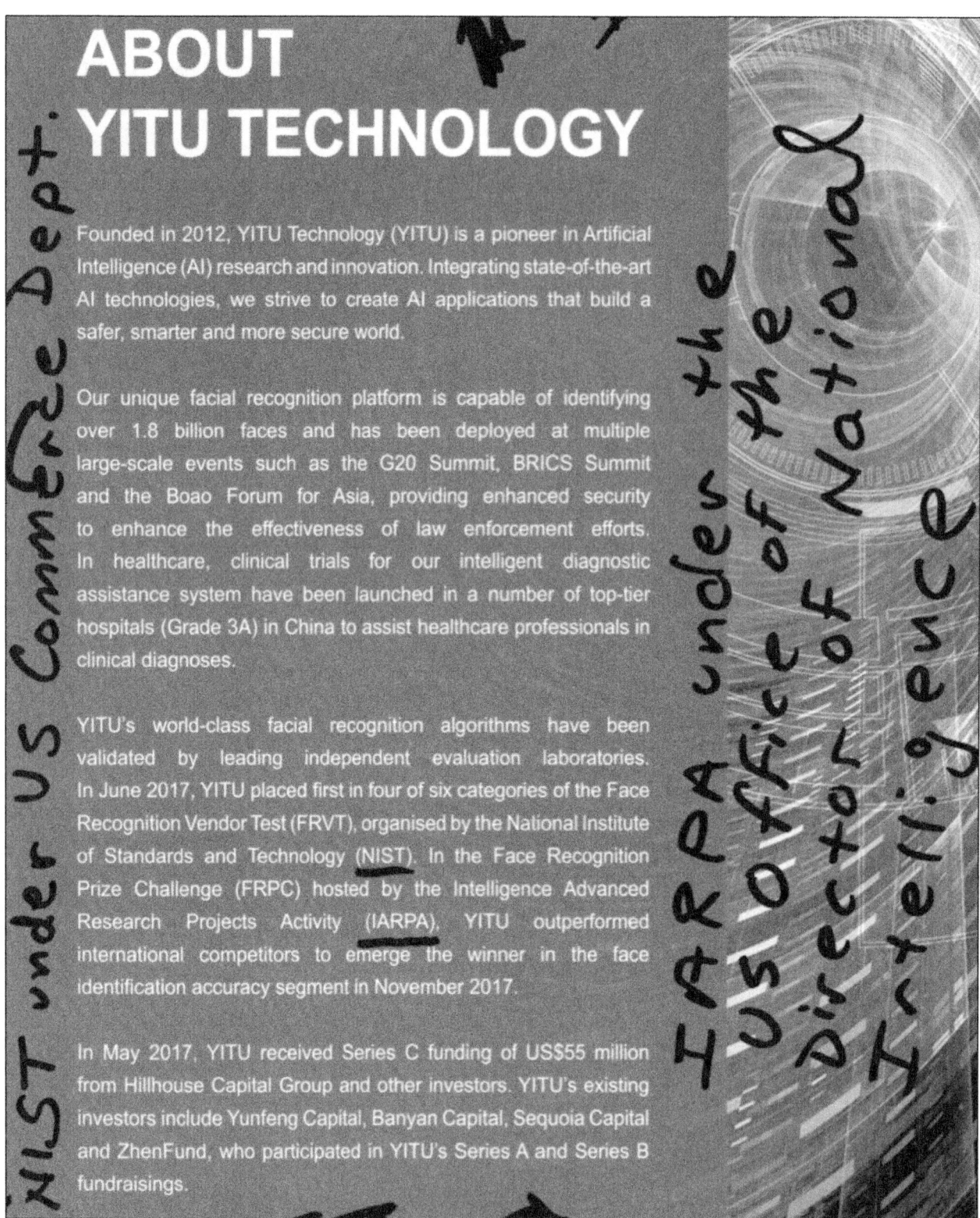

YITU Brochure. 2018 Defense Services Asia. (2/22).

WORLD-CLASS ARTIFICIAL INTELLIGENCE ALGORITHMS

RANKED FIRST IN THE FACE RECOGNTION VENDOR TEST AND WINNER OF THE FACE RECOGNITION PRIZE CHALLENGE

YITU RANKED FIRST IN THE FACE RECOGNITION VENDOR TEST

Organised by the National Institute of Standards and Technology (NIST), the Face Recognition Vendor Test (FRVT) is one of the world's most stringent industry benchmarks, with real-world scenarios provided by the U.S. Department of Homeland Security. The FRVT also provides official guidance for U.S. government purchases.

YITU ranked first across four categories, with an accuracy rating of 95.5%. The result was 2.0% higher than the second best placing, and the best ever result since 2014.

YITU's win also broke down barriers by being the first Chinese tech firm to win the FRVT.

WINNER OF THE FACE RECOGNITION PRIZE CHALLENGE

Hosted by the Intelligence Advanced Research Projects Activity (IARPA) under the U.S. Office of the Director of National Intelligence, the Face Recognition Prize Challenge (FRPC) recognises developers with the most accurate unconstrained face recognition algorithm.

YITU outperformed over a dozen international competitors to take first place in the face identification accuracy segment of the FRPC where YITU's facial recognition software was found most accurate for the 1:N matching accuracy segment.

The win is a recognition of YITU's AI algorithms, which have the ability to recognise faces from images complicated by environmental conditions such as non-frontal head poses, poor and uneven illumination, motion blur and non-neutral facial expressions.

YITU Brochure. 2018 Defense Services Asia. (3/22).

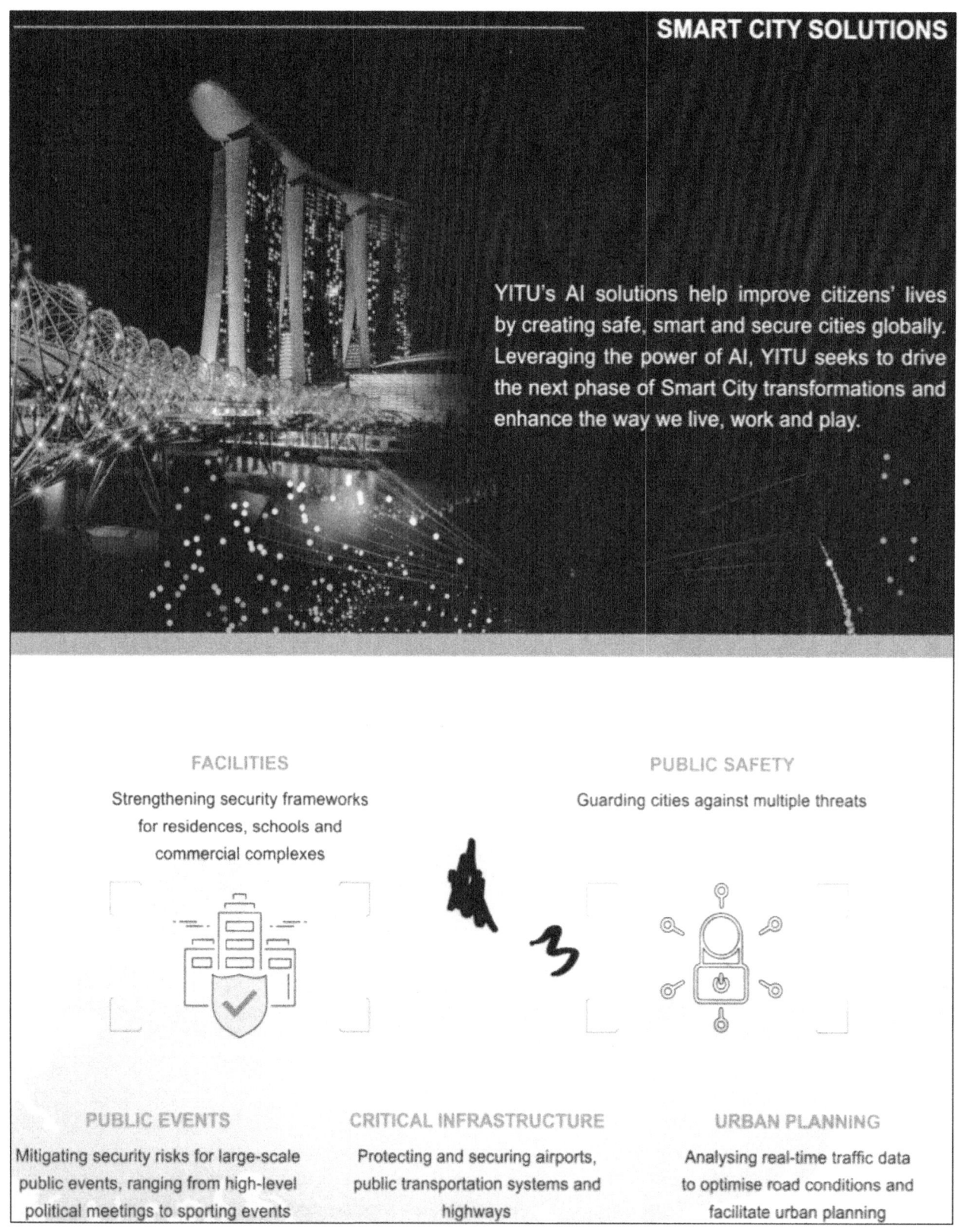

YITU Brochure. 2018 Defense Services Asia. (4/22).

SMART SECURITY SOLUTIONS

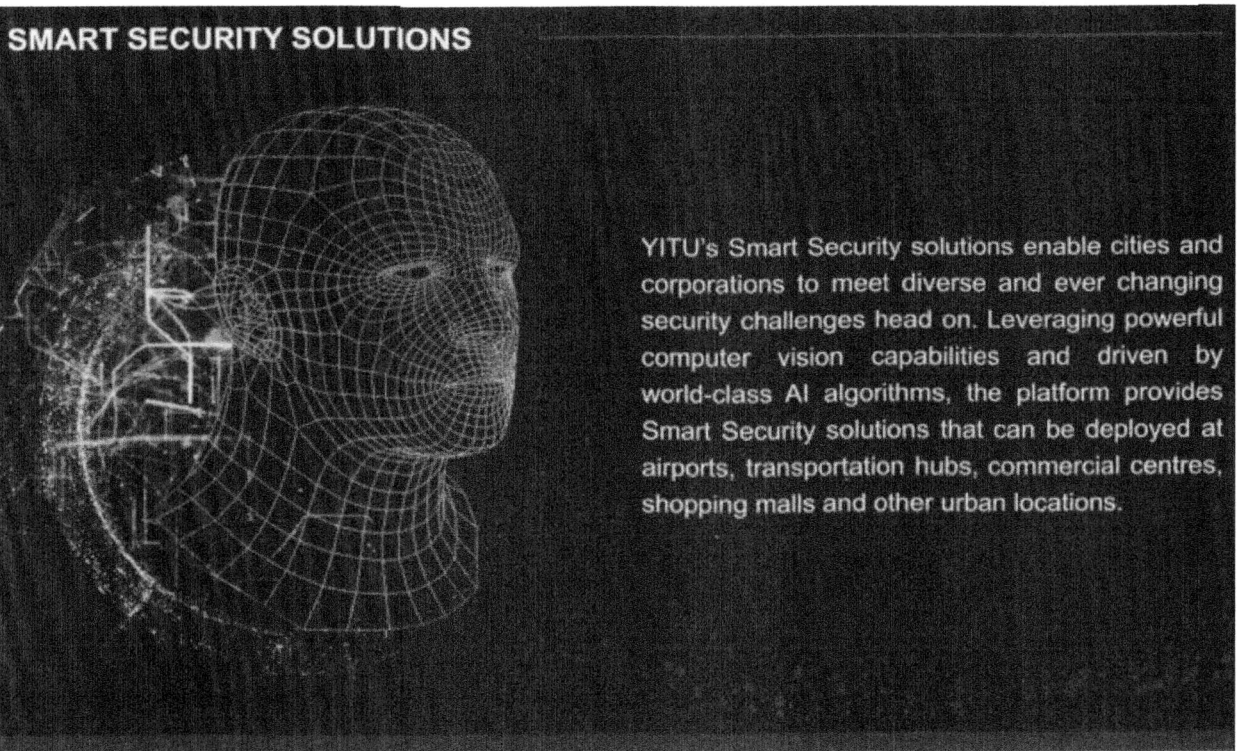

YITU's Smart Security solutions enable cities and corporations to meet diverse and ever changing security challenges head on. Leveraging powerful computer vision capabilities and driven by world-class AI algorithms, the platform provides Smart Security solutions that can be deployed at airports, transportation hubs, commercial centres, shopping malls and other urban locations.

YITU's proprietary facial recognition platform brings together advanced machine vision algorithms, high-performance distributed computation and storage, massive operations, maintenance and other components. It is the world's largest portrait comparison platform, and is capable of identifying over 1.8 billion individuals within seconds. Providing intelligent urban security solutions for cities, it is presently used by 150 municipal public security systems and 20 provincial public security departments across China.

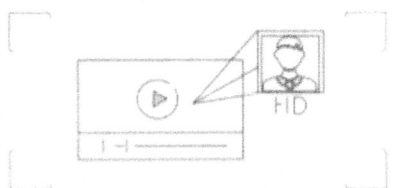 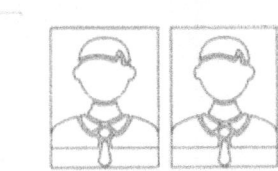  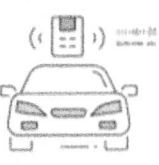

DYNAMIC PORTRAIT SYSTEM

YITU's Dynamic Portrait System supports most mainstream video stream standards in the industry. Incorporating facial recognition and tracking, and real-time deployment and control, the system promotes public safety at crowded areas such as commercial and shopping zones, train stations, airports, and residential districts.

STATIC PORTRAIT SYSTEM

Available for 1:1, 1:N and n:N comparison, YITU's Static Portrait System supports large-scale portrait comparison and has been widely used in the security and financial sectors. It can be used to build a database for urban populations to quickly and accurately identify persons of interest.

VEHICLE RECOGNITION SYSTEM

Beyond facial recognition, YITU's powerful AI algorithms have also been implemented in YITU's intelligent vehicle recognition platform, the first-of-its-kind in China. Deploying YITU's proprietary technologies of car brand identification, false license plate analysis and car location tracking based on a vehicle's visual features, the platform assists law enforcement officers in detecting the use of false license plates and monitoring real time vehicle entries and exits.

YITU Brochure. 2018 Defense Services Asia. (5/22).

SMART CASINO SOLUTIONS

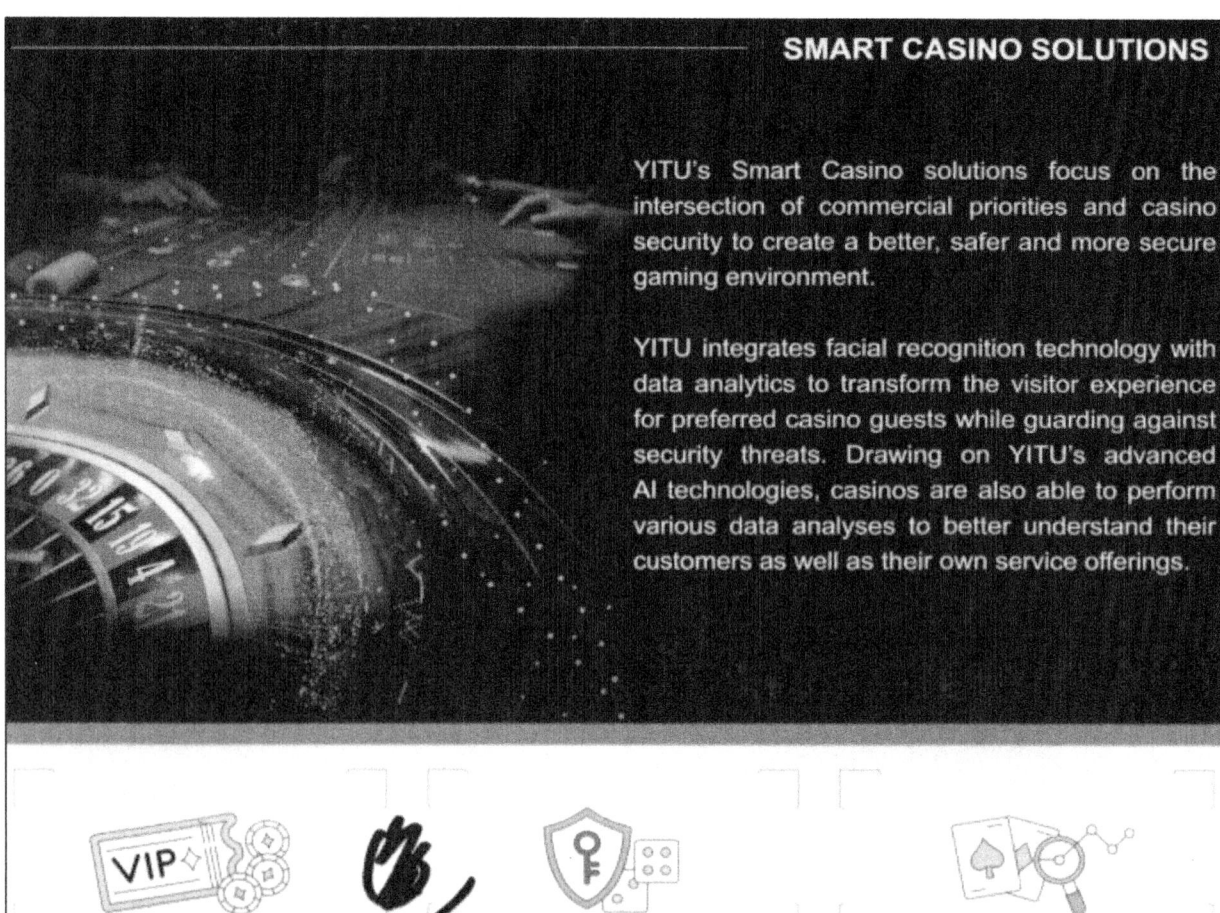

YITU's Smart Casino solutions focus on the intersection of commercial priorities and casino security to create a better, safer and more secure gaming environment.

YITU integrates facial recognition technology with data analytics to transform the visitor experience for preferred casino guests while guarding against security threats. Drawing on YITU's advanced AI technologies, casinos are also able to perform various data analyses to better understand their customers as well as their own service offerings.

VIP GUEST RECOGNITION

With YITU's unique facial recognition system, casinos can promptly identify and greet VIP guests the moment they enter the casino. Instant alerts sent to relationship managers and hospitality staff provide actionable insights on VIP guests, from their gaming preferences to even their choice of refreshments. This allows casinos to customise and provide a positive gaming experience for these guests, such as ushering them to their favourite card table or serving their preferred beverage.

ENHANCED CASINO SECURITY

Seamlessly integrated with existing casino security systems, YITU's facial recognition system detects faces from live video streams and compares them against photos of unauthorised individuals from a casino's "blacklist" database.

Whether these are individuals who have applied for a casino self-exclusion scheme or known card counters and advantage players barred from gaming establishments, YITU's system is capable of detecting these individuals in real-time and alerting security staff to their presence in the casino. This not only helps casinos prevent potential losses but also creates a safer casino environment for staff and patrons.

IMPROVED CASINO OFFERINGS

YITU enables casinos with powerful insights to optimise their service offerings through our proprietary path tracking technology. In order to better understand guests' behaviours and preferences, casinos can track guest movements not just within the casino but also any attached retail or dining establishments. Insights generated can help shape a casino's promotional activities to yield better guest experiences.

By analysing guests' interactions with the casino games available, casinos can also ascertain guests' gaming preferences and optimise the gaming mix as appropriate. YITU's data analysis helps casinos stay competitive in the fast-paced casino industry where guest experience and a superior games selection is key.

YITU Brochure. 2018 Defense Services Asia. (6/22).

SMART HEALTHCARE SOLUTIONS

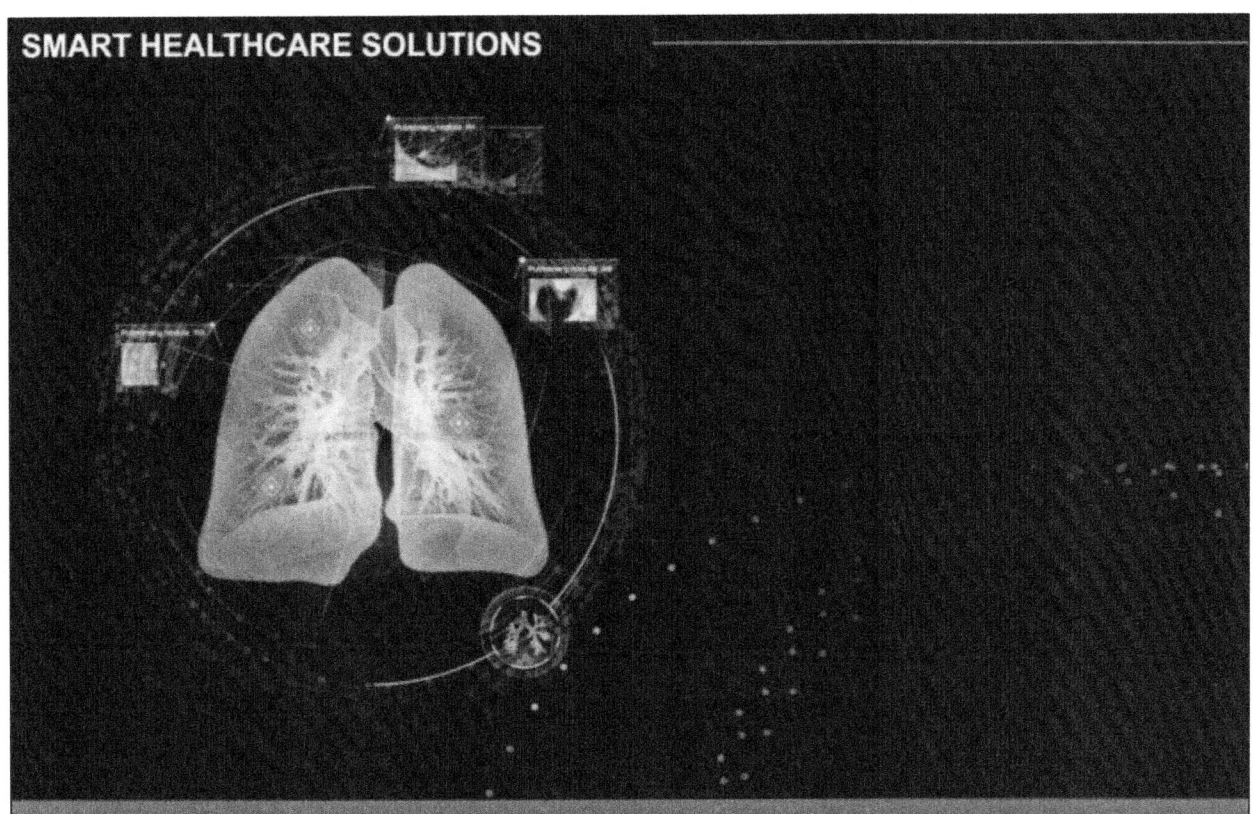

OUR AICARE® RANGE

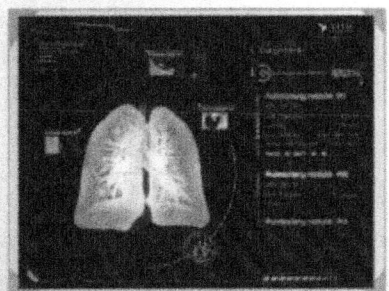
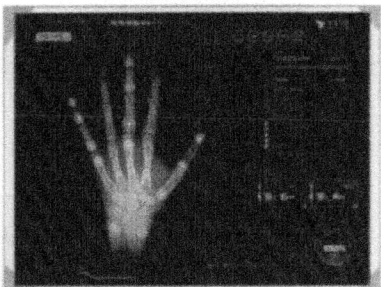
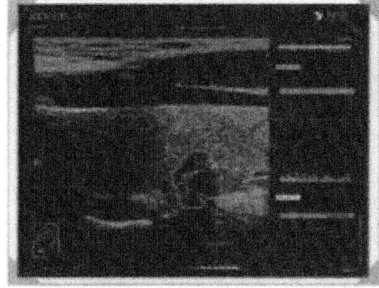

INTELLIGENT AUXILIARY DIAGNOSIS ON CHEST CT IMAGE

Based on deep learning technology, this system can read chest CT images within seconds, analyse sizes, characters and anatomical locations of lesions, diagnose whether they are benign or malignant by combining clinical information, and provide structured diagnosis reports.

INTELLIGENT AUXILIARY DIAGNOSIS ON CHILDREN'S SKELETAL AGE

Based on deep learning technology and Tanner-Whitehouse (TW3) standards of bone age measurement, this system can automatically calculate children's bone age with precision of +/- one month compared to experts' diagnosis.

ULTRASONIC INTELLIGENT AUXILIARY DIAGNOSIS

Based on deep learning technology and doctors' clinical diagnosis experience, this product can automatically detect different types of ultrasonic images of mammary glands, thyroid glands and other organs, and provide diagnosis reports for lesion classification.

YITU Brochure. 2018 Defense Services Asia. (7/22).

YITU's healthcare team of AI scientists, big data experts and senior medical consultants are focused on developing Smart Healthcare solutions that meet evolving healthcare needs. YITU has applied AI technologies to large volumes of medical data such as medical imaging, medical records, genetics, medical tests and research to create an integrated platform for medical data analysis, including data ETL (extract, transform and load), data annotation, model training and clinical workflow optimisation.

Apart from providing assisted diagnosis systems and data analysis platforms, YITU also partners leading 3A hospitals in China on scientific research projects. Through the interdisciplinary integration of medicine and other scientific fields, YITU seeks to improve healthcare outcomes and provide better results for patients with our AICARE® range of Smart Healthcare solutions.

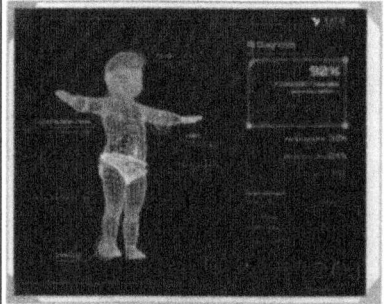 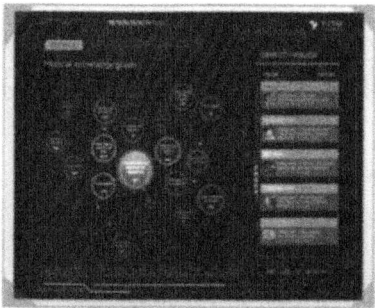 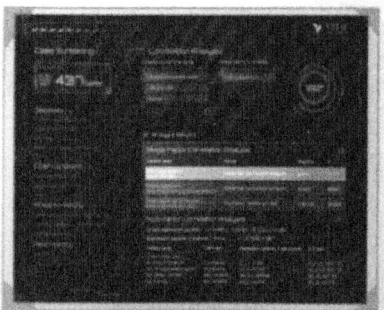

INTELLIGENT AUXILIARY DIAGNOSIS ON PEDIATRIC OUTPATIENTS

This is a pediatric clinical-assisted diagnosis product based on pediatric experts' experience, mass clinical medical records and advanced deep learning technology.

INTELLIGENT MEDICAL RECORD SEARCH ENGINE

Based on advanced deep learning and NLP technology, this is an intelligent platform for searching, viewing and analysing electronic medical records.

CLINICAL INTELLIGENT SCIENTIFIC RESEARCH PLATFORM

This assisted scientific research platform covers the entire research work flow, including clinical data preparation, data processing and data analysis.

YITU Brochure. 2018 Defense Services Asia. (8/22).

SMART FINANCE SOLUTIONS

Led by rapid advancements in AI, banks and financial services providers are increasingly shifting away from traditional financial services to financial technology. YITU's AI solutions help financial institutions transform, streamline and improve their financial processes and operational efficiency.

Offering both retail and institutional applications across industries such as banking, internet finance, insurance, securities finance, fund management and auto finance, YITU's full range of Smart Finance solutions are customised to meet future business needs, now.

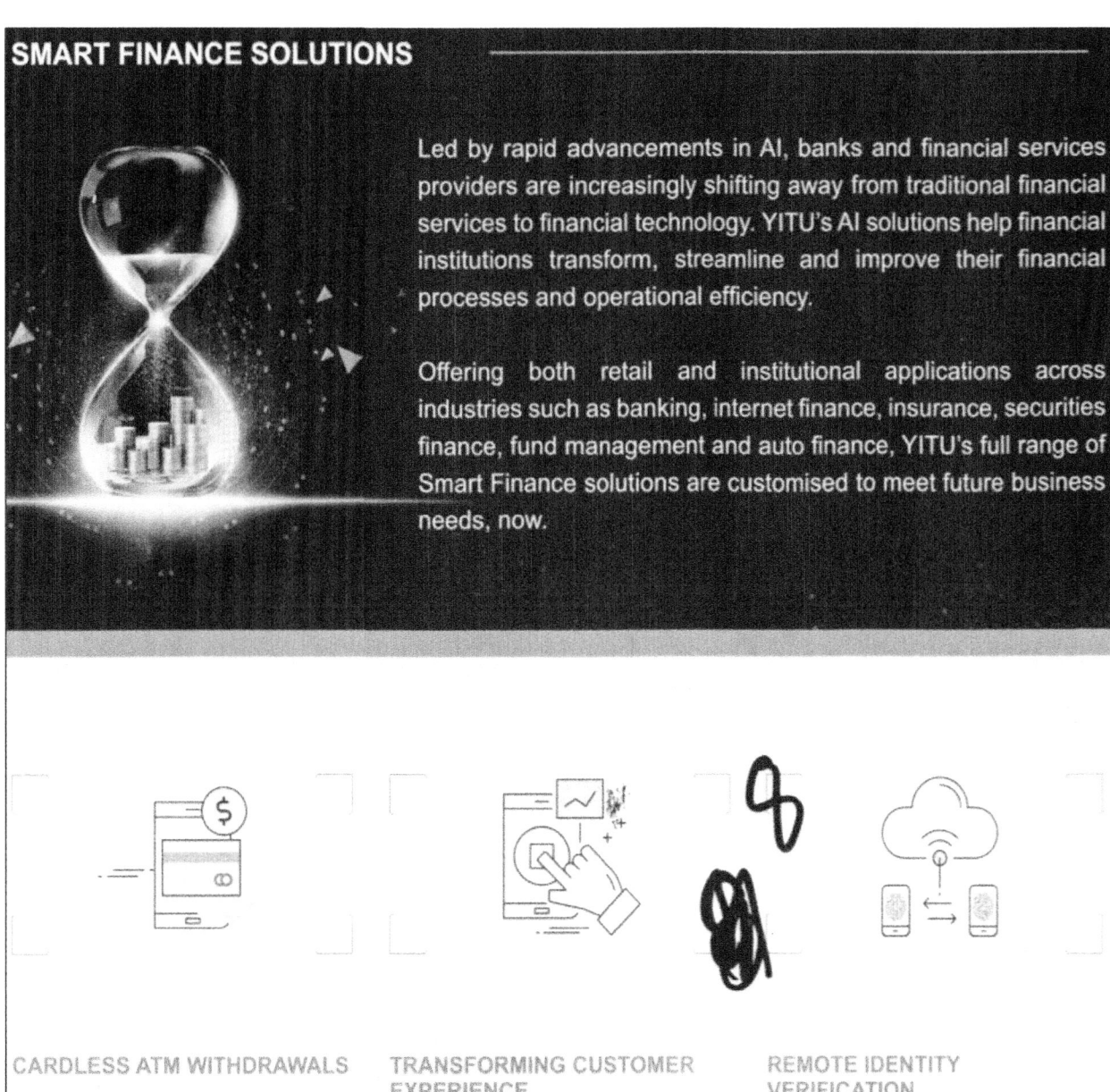

CARDLESS ATM WITHDRAWALS

A faster, more secure way to withdraw cash from an ATM, YITU's facial recognition system has been rolled out across China Merchant Bank's network of 1,500 ATMs in China. YITU's proprietary binocular liveness detection and facial recognition software offer convenient and secure face scanning and identity verification.

TRANSFORMING CUSTOMER EXPERIENCE

Leveraging YITU's machine learning and AI solutions, financial institutions are able to analyse bank branch data and video footage for insights on how to enhance the customer experience.

REMOTE IDENTITY VERIFICATION

YITU offers businesses a quick and reliable ID verification process to verify the identities of digital customers. Using liveness detection, facial recognition, and ID scanning and authentication, YITU's platform enables secure transactions over mobile channels.

YITU Brochure. 2018 Defense Services Asia. (9/22).

SMART HARDWARE

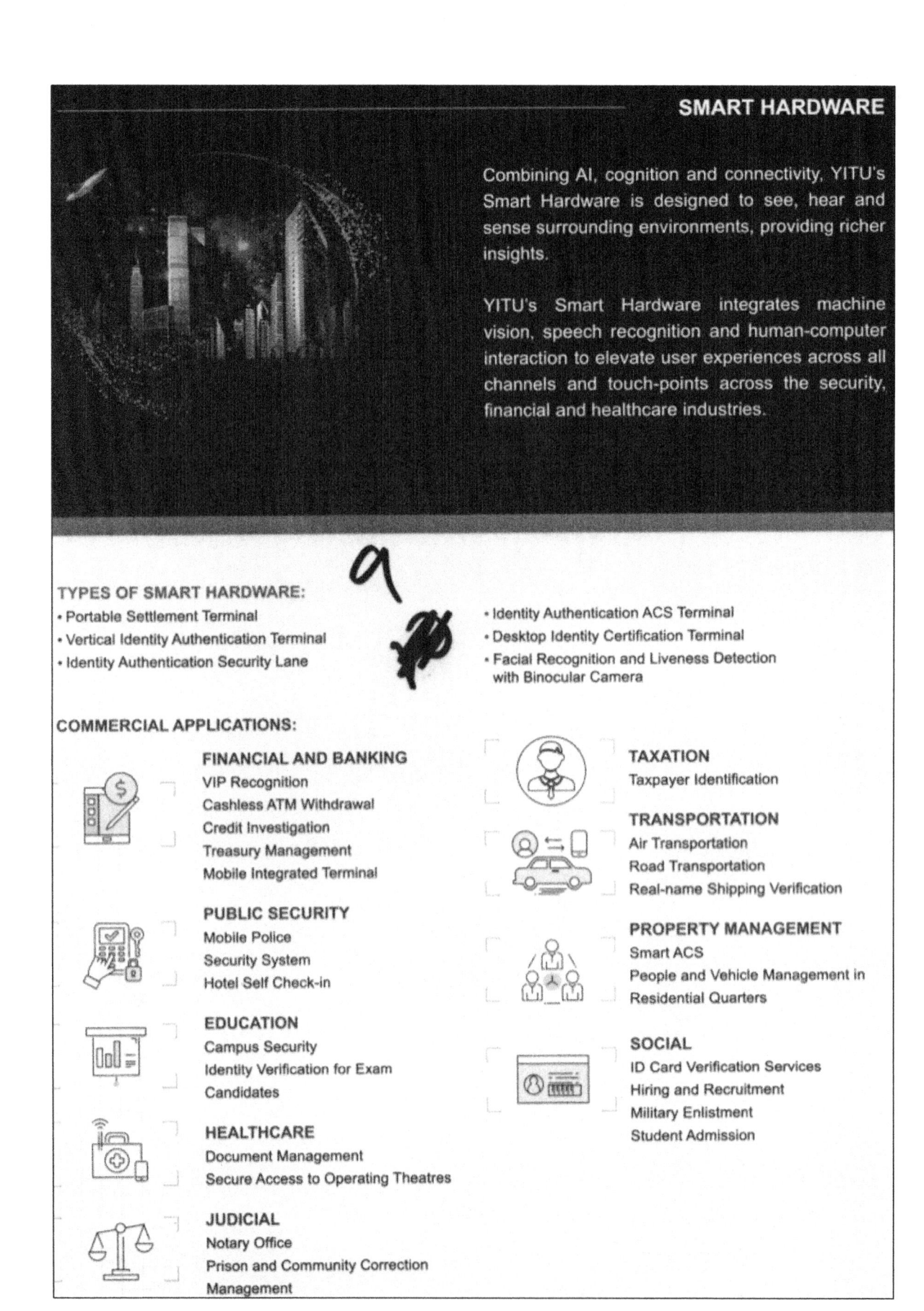

Combining AI, cognition and connectivity, YITU's Smart Hardware is designed to see, hear and sense surrounding environments, providing richer insights.

YITU's Smart Hardware integrates machine vision, speech recognition and human-computer interaction to elevate user experiences across all channels and touch-points across the security, financial and healthcare industries.

TYPES OF SMART HARDWARE:
- Portable Settlement Terminal
- Vertical Identity Authentication Terminal
- Identity Authentication Security Lane
- Identity Authentication ACS Terminal
- Desktop Identity Certification Terminal
- Facial Recognition and Liveness Detection with Binocular Camera

COMMERCIAL APPLICATIONS:

FINANCIAL AND BANKING
VIP Recognition
Cashless ATM Withdrawal
Credit Investigation
Treasury Management
Mobile Integrated Terminal

PUBLIC SECURITY
Mobile Police
Security System
Hotel Self Check-in

EDUCATION
Campus Security
Identity Verification for Exam Candidates

HEALTHCARE
Document Management
Secure Access to Operating Theatres

JUDICIAL
Notary Office
Prison and Community Correction Management

TAXATION
Taxpayer Identification

TRANSPORTATION
Air Transportation
Road Transportation
Real-name Shipping Verification

PROPERTY MANAGEMENT
Smart ACS
People and Vehicle Management in Residential Quarters

SOCIAL
ID Card Verification Services
Hiring and Recruitment
Military Enlistment
Student Admission

YITU Brochure. 2018 Defense Services Asia. (10/22).

FOUNDERS

LEO ZHU
CO-FOUNDER & CEO, YITU TECHNOLOGY

Leo Zhu is the Co-Founder and CEO of YITU Technology. He received his PhD in Statistics from the University of California, Los Angeles (UCLA), and was a student of Professor Alan Yuille, a disciple of Stephen William Hawking, specialising in statistical modeling of computer vision and AI.

Leo was a post-doctoral fellow at Massachusetts Institute of Technology's (MIT) AI laboratory, specialising in the study of brain science and computational photography. He was also a research fellow at the Courant Institute of Mathematical Sciences at New York University (NYU), helmed by Yann Lecun, who is renowned for being the founder of deep learning.

LIN CHENXI
CO-FOUNDER, YITU TECHNOLOGY

Lin Chenxi is the Co-Founder of YITU Technology. In his former role as senior expert at Alibaba Cloud, Lin led a team of over a hundred senior engineers to set up the Apsara distributed cloud computing operating system, the largest of its kind with independent intellectual property rights in China.

Before joining the Alibaba Group, Lin was with Microsoft Research Asia (MSRA), specialising in the fields of machine learning, computer vision, information retrieval and distributed systems.

Lin received his Masters degree from Shanghai Jiao Tong University in 2005. In 2002, Lin was selected as a member of the ACM International Collegiate Programming Contest (ACM-ICPC) team, and the team emerged as champion of the ACM-ICPC, marking an inaugural win by a team from Asia. Lin was also an outstanding student, having received the first Chancellor of Shanghai Jiao Tong University Award in 2003.

YITU Brochure. 2018 Defense Services Asia. (11/22).

LEO ZHU
CO-FOUNDER & CEO, YITU TECHNOLOGY

Leo Zhu is the Co-Founder and CEO of YITU Technology. He received his PhD in Statistics from the University of California, Los Angeles (UCLA), and was a student of Professor Alan Yuille, a disciple of Stephen William Hawking, specialising in statistical modeling of computer vision and AI.

Leo was a post-doctoral fellow at Massachusetts Institute of Technology's (MIT) AI laboratory, specialising in the study of brain science and computational photography. He was also a research fellow at the Courant Institute of Mathematical Sciences at New York University (NYU), helmed by Yann Lecun, who is renowned for being the founder of deep learning.

Close-Up: YITU Brochure. 2018 Defense Services Asia. (12/22). Close-up.

LIN CHENXI
CO-FOUNDER, YITU TECHNOLOGY

Lin Chenxi is the Co-Founder of YITU Technology. In his former role as senior expert at Alibaba Cloud, Lin led a team of over a hundred senior engineers to set up the Apsara distributed cloud computing operating system, the largest of its kind with independent intellectual property rights in China.

Before joining the Alibaba Group, Lin was with Microsoft Research Asia (MSRA), specialising in the fields of machine learning, computer vision, information retrieval and distributed systems.

Lin received his Masters degree from Shanghai Jiao Tong University in 2005. In 2002, Lin was selected as a member of the ACM International Collegiate Programming Contest (ACM-ICPC) team, and the team emerged as champion of the ACM-ICPC, marking an inaugural win by a team from Asia. Lin was also an outstanding student, having received the first Chancellor of Shanghai Jiao Tong University Award in 2003.

Close-Up: YITU Brochure. 2018 Defense Services Asia. (13/22). Close-up.

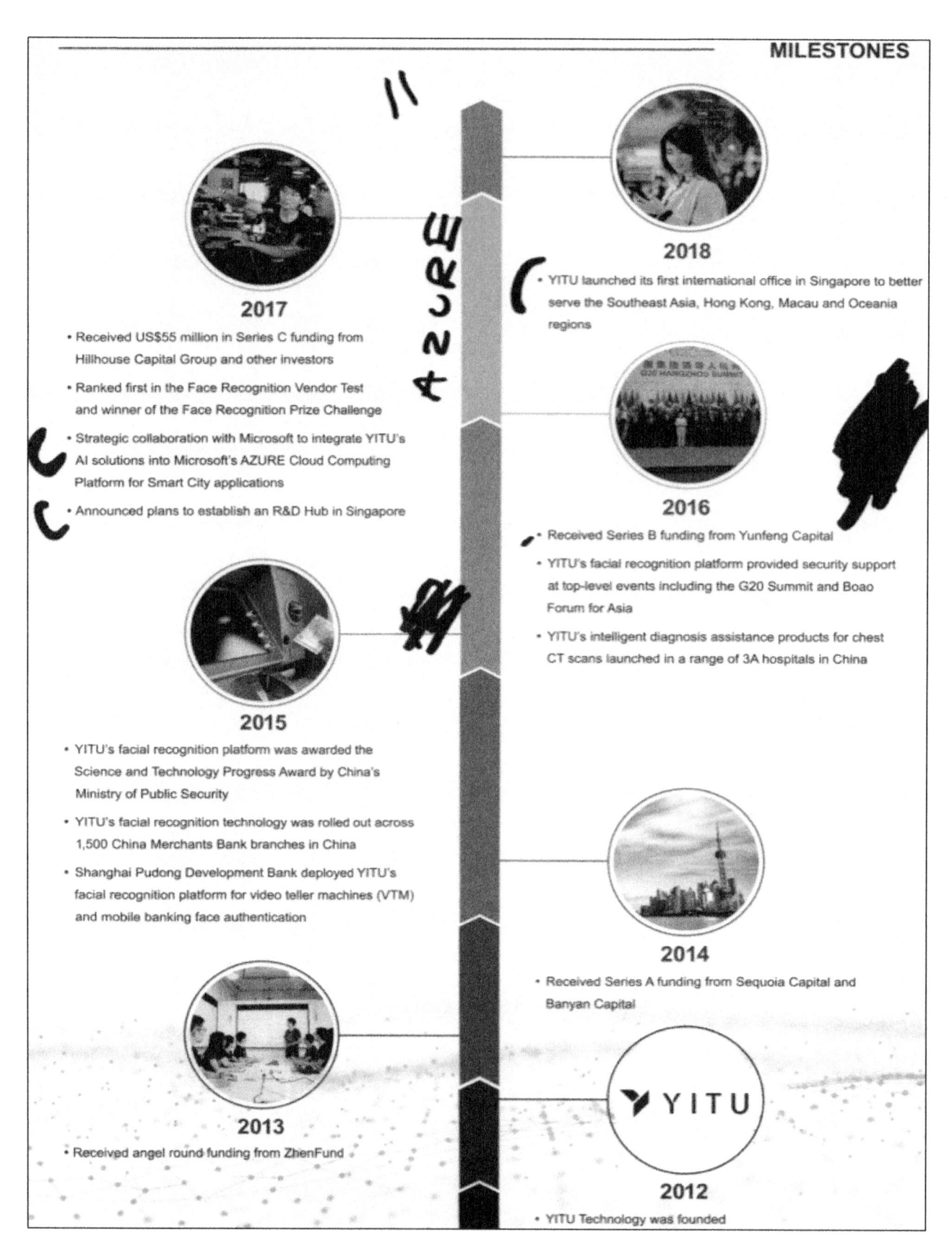

MILESTONES

2018
- YITU launched its first international office in Singapore to better serve the Southeast Asia, Hong Kong, Macau and Oceania regions

2017
- Received US$55 million in Series C funding from Hillhouse Capital Group and other investors
- Ranked first in the Face Recognition Vendor Test and winner of the Face Recognition Prize Challenge
- Strategic collaboration with Microsoft to integrate YITU's AI solutions into Microsoft's AZURE Cloud Computing Platform for Smart City applications
- Announced plans to establish an R&D Hub in Singapore

2016
- Received Series B funding from Yunfeng Capital
- YITU's facial recognition platform provided security support at top-level events including the G20 Summit and Boao Forum for Asia
- YITU's intelligent diagnosis assistance products for chest CT scans launched in a range of 3A hospitals in China

2015
- YITU's facial recognition platform was awarded the Science and Technology Progress Award by China's Ministry of Public Security
- YITU's facial recognition technology was rolled out across 1,500 China Merchants Bank branches in China
- Shanghai Pudong Development Bank deployed YITU's facial recognition platform for video teller machines (VTM) and mobile banking face authentication

2014
- Received Series A funding from Sequoia Capital and Banyan Capital

2013
- Received angel round funding from ZhenFund

2012
- YITU Technology was founded

YITU Brochure. 2018 Defense Services Asia. (14/22).

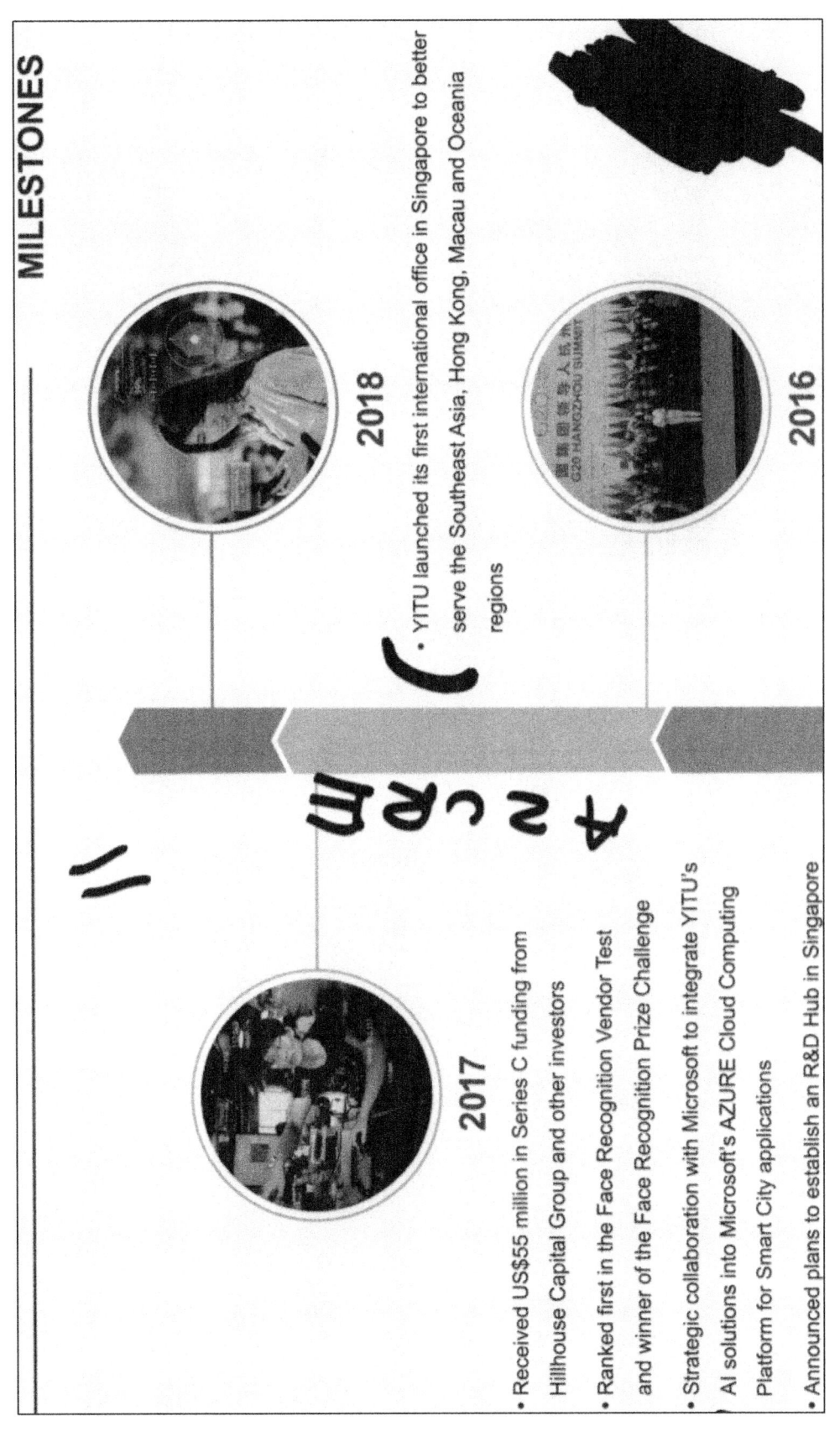

MILESTONES

2018
- YITU launched its first international office in Singapore to better serve the Southeast Asia, Hong Kong, Macau and Oceania regions

2016

2017
- Received US$55 million in Series C funding from Hillhouse Capital Group and other investors
- Ranked first in the Face Recognition Vendor Test and winner of the Face Recognition Prize Challenge
- Strategic collaboration with Microsoft to integrate YITU's AI solutions into Microsoft's AZURE Cloud Computing Platform for Smart City applications
- Announced plans to establish an R&D Hub in Singapore

Close-Up: YITU Brochure. 2018 Defense Services Asia. (15/22).

2016

- Received Series B funding from Yunfeng Capital

- YITU's facial recognition platform provided security support at top-level events including the G20 Summit and Boao Forum for Asia

- YITU's intelligent diagnosis assistance products for chest CT scans launched in a range of 3A hospitals in China

Close-Up: YITU Brochure. 2018 Defense Services Asia. (16/22).

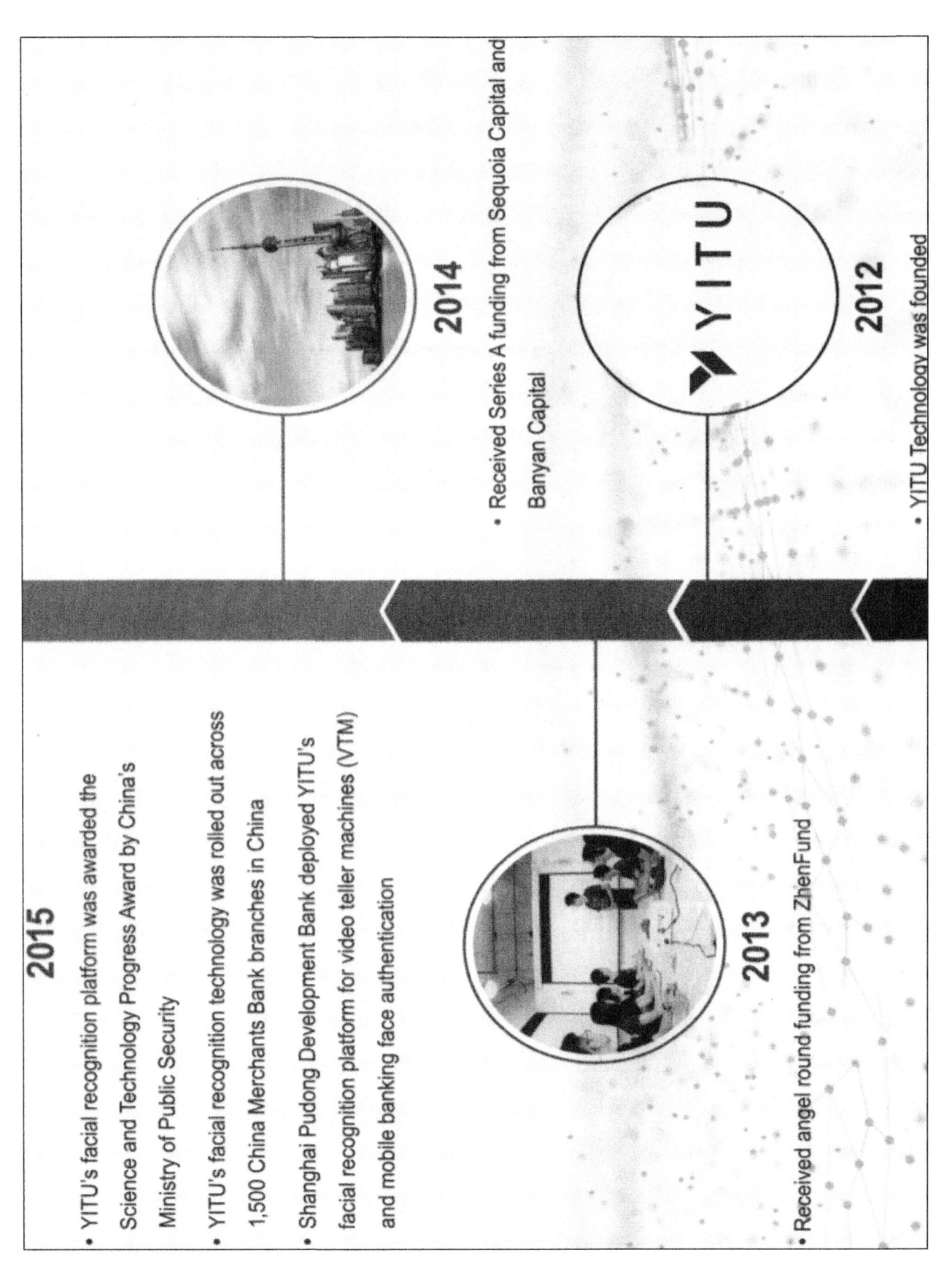

Close-Up: YITU Brochure. 2018 Defense Services Asia. (17/22).

2012
- YITU Technology was founded

2013
- Received angel round funding from ZhenFund

2014
- Received Series A funding from Sequoia Capital and Banyan Capital

2015
- YITU's facial recognition platform was awarded the Science and Technology Progress Award by China's Ministry of Public Security
- YITU's facial recognition technology was rolled out across 1,500 China Merchants Bank branches in China
- Shanghai Pudong Development Bank deployed YITU's facial recognition platform for video teller machines (VTM) and mobile banking face authentication

YITU Brochure. 2018 Defense Services Asia. (18/22).

Close-Up: YITU Brochure. 2018 Defense Services Asia. (19/22).

Close-Up: YITU Brochure. 2018 Defense Services Asia. (20/22).

YITU

YITU Singapore
8 Marina View, #32-06, Asia Square Tower 1, Singapore 018960
T: +65 6631 8988   F: +65 6631 8990   W: www.yitutech.sg   E: YITU-Singapore@yitu-inc.com

YITU Silicon Valley
101 Jefferson Dr, Menlo Park, CA 94025

YITU Shanghai
23F, Tower 1, Shanghai Jin Hongqiao International Center, 523 Loushanguan Road,
Changning District, Shanghai, China
W: www.yitutech.com

YITU Beijing
601C, Tower B, Qinghua International Technology Communication Center,
8 Qinghua Science and Technology Park, 1 Zhongguncun East Rd, Haidian District, Beijing, China
W: www.yitutech.com

YITU Hangzhou
406, Building 4, Zhejiang Fortune Center, 83 Gudun Rd, Xihu District, Hangzhou, China
W: www.yitutech.com

YITU Brochure. 2018 Defense Services Asia. (21/22).

SOUTHEAST ASIA, HONG KONG AND MACAU

- <u>bUILd</u>
- <u>Certis</u>: formerly Certis CISCO Security Private Limited and CISCO Security Private Limited, is one of the five commercial Auxiliary Police forces authorized to provide armed security officers to government organizations and private companies in Singapore.
- <u>Changi Airport Group</u> (Singapore)
- <u>Chubb</u> (United Technologies)
- <u>Milestone</u>: Milestone Systems. a company in the Canon group, is a global industry leader in open platform IP video management software.
- <u>Auxiliary Force SDN BHD</u>: Security systems services, except locksmithing (Malaysia).
- <u>VST ECS</u>: VST ECS Phils. Inc. is one of the leading Information and Communications Technology (ICT) distributors in the Philippines.

CHINA

- <u>China Immigration Inspection</u> (中国边检): Chinese border patrol is the government agency responsible for controlling its borders.
- <u>China Customs</u>: General Administration of Customs.
- <u>China Unionpay</u>
- <u>China Merchants Bank</u>
- <u>Wanda Group</u>
- <u>Fudan University Shanghai Cancer Center</u>
- <u>Microsoft</u>
- <u>Huawei</u>
- <u>Tencent</u>

Editor Note with Explanations relating to YITU Brochure. 2018 Defense Services Asia. (22/22).

LANU-M1 VEHICLE MOUNTED UAV JAMMER

2018 DSA

EW

LANU-M1 Vehicle Mounted UAV Jammer. Poly Technologies. 2018 Defence Services Asia (Malaysia). (1/3).

GENERAL DESCRIPTION

LANU-M1 is a vehicle mounted UAV jammer which can created a 360° protection area where can be used in sudden attack, vital important site protection, VIP convoy and ammunition vehicle convoy. It is very easy to be installed on the vehicle.

SPECIFICATION

	LANU-M1
Frequency Band	Band1: 2400-2485Mhz
	Band2: 5725-5850Mhz
	Band3: 1559-1620Mhz
EIRP	Band1: 48dBm(64w)
	Band2: 43dBm(20w)
	Band3: 45dBm(32w)
Wave Beam Width	Band1: Horizontal360° Vertical75°
	Band2: Horizontal360° Vertical75°
	Band3: Horizontal360° Vertical90°
Jamming Range	Band1: ≥ 1000m
	Band2: ≥ 1000m
	Band3: ≥ 2000m
High Temperature Alarm	√
Low Output Power Alarm	√
Power Consumption	≤ 400w
Voltage	DC48-60V
Working Mode	Continuous
Weight	16kg
Size	500mm×250mm×190mm
Water Proof Level	IP68
Working Temperature	Main Host: -40°C ~ +60°C

LANU-M1 Vehicle MOUNTED UAV Jammer. Poly Technologies. 2018 Defence Services Asia (Malaysia). (2/3).

SPECIFICATION

	LANU-M1
Frequency Band	Band1: 2400-2485Mhz
	Band2: 5725-5850Mhz
	Band3: 1559-1620Mhz
EIRP	Band1: 48dBm(64w)
	Band2: 43dBm(20w)
	Band3: 45dBm(32w)
Wave Beam Width	Band1: Horizontal360° Vertical75°
	Band2: Horizontal360° Vertical75°
	Band3: Horizontal360° Vertical90°
Jamming Range	Band1: ≥ 1000m
	Band2: ≥ 1000m
	Band3: ≥ 2000m
High Temperature Alarm	√
Low Output Power Alarm	√
Power Consumption	≤ 400w
Voltage	DC48-60V
Working Mode	Continuous
Weight	16kg
Size	500mm×250mm×190mm
Water Proof Level	IP68
Working Temperature	Main Host: -40℃ ~ +60℃

CLOSE-UP. LANU-M1 Vehicle MOUNTED UAV Jammer. Poly Technologies. 2018 Defence Services Asia (Malaysia). (3/3).

LANU-W1 Portable WIDE Angle UAV Jammer. Poly Technologies. 2018 Defence Services Asia (Malaysia). (1/4).

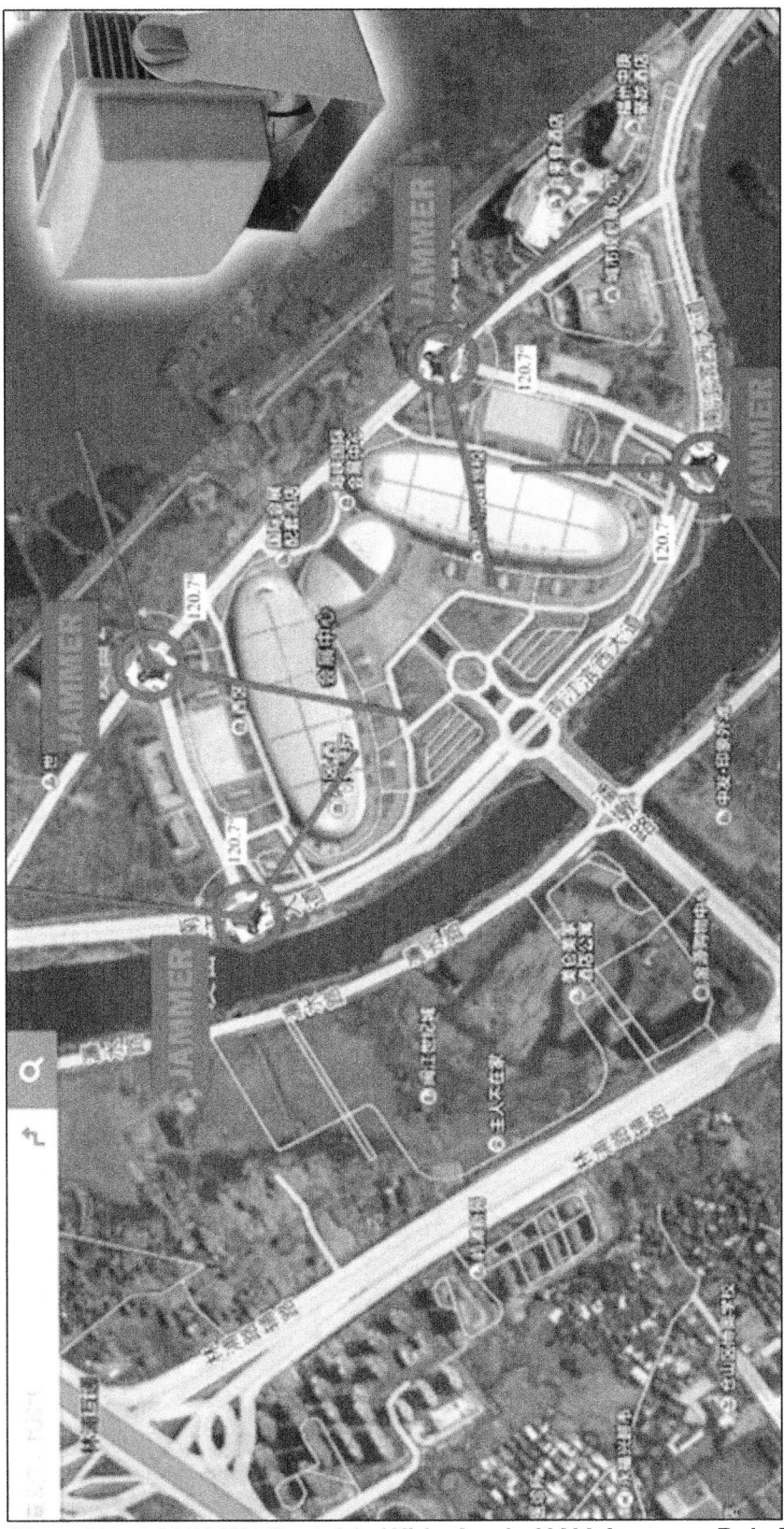

Close-Up. LANU-W1 Portable Wide Angle UAV Jammer. Poly Technologies. 2018 Defence Services Asia (Malaysia). (2/4).

GENERAL DESCRIPTION

LANU-W1 is a high efficiency EW (Electronic Warfare) device aiming to intercept civilian UAVs. LANU-W1 employs advanced DDS and MMIC technologies and is effective in jamming and interfering with control &communication channels of most civilian UAVs currently available on the market. The UAV jammer is designed to force land or turn away one or more invading UAVs. The portable design makes it easy to be installed.It has been widely used in military bases, prisons, vital sits and airpots, etc.

SPECIFICATION

	LANU-W1
Frequency Band	Band1: 2400-2485Mhz
	Band2: 5725-5850Mhz
	Band3: 1559-1620Mhz
	Band4: 915-930MHz depends on customer's requirement
Power	Band1: 45dBm(32w)
	Band2: 41dBm(13w)
	Band3: 45dBm(32w)
Wave Beam Width	Band1: Horizontal120°　Vertical120°
	Band2: Horizontal120°　Vertical120°
	Band3: Horizontal150°　Vertical120°
Jamming Range	Band1: ≥ 2000m
	Band2: ≥ 2000m
	Band3: ≥ 3000m
Power Consumption	350w
Remote Control Distance	≥100m (remote control), ≥1000m (combined use with Repeaters)
Voltage	DC100-240V or 50-60Hz
Working Mode	Continuous
Weight	10.5kg
Size	300mm×390mm×230mm
Water Proof Level	IP67
Working Temperature	Main Host: -40℃ ~ +55℃

LANU-W1 Portable Wide Angle UAV Jammer. Poly Technologies. 2018 Defence Services Asia (Malaysia). (3/4).

SPECIFICATION

	LANU-W1
Frequency Band	Band1: 2400-2485Mhz
	Band2: 5725-5850Mhz
	Band3: 1559-1620Mhz
	Band4: 915-930MHz depends on customer's requirement
Power	Band1: 45dBm(32w)
	Band2: 41dBm(13w)
	Band3: 45dBm(32w)
Wave Beam Width	Band1: Horizontal120° Vertical120°
	Band2: Horizontal120° Vertical120°
	Band3: Horizontal150° Vertical120°
Jamming Range	Band1: ≥ 2000m
	Band2: ≥ 2000m
	Band3: ≥ 3000m
Power Consumption	350w
Remote Control Distance	≥100m (remote control), ≥1000m (combined use with Repeaters)
Voltage	DC100-240V or 50-60Hz
Working Mode	Continuous
Weight	10.5kg
Size	300mm×390mm×230mm
Water Proof Level	IP67
Working Temperature	Main Host: -40℃ ~ +55℃

Close-Up. LANU-W1 Portable Wide Angle UAV Jammer. Poly Technologies. 2018 Defence Services Asia (Malaysia). (4/4).

"寂静狩猎者"低小慢目标激光拦截系统

推介关键词：

硬杀伤：3万瓦激光器，直接毁伤飞行器

多平台：可采用移动式和固定式的使用模式

所属我司本次推介系统：

反恐系统　　　　　　　　要地防御系统、重大活动保障系统

简介：

寂静狩猎者低小慢目标激光拦截系统由指挥控制分系统、雷达搜索分系统、光电跟踪分系统和激光发射器分系统组成，能够实现目标搜索、捕获、定位和击毁的作战功能，该系统可部署在楼宇顶端等重要区域。

指标

功率输出	5kw	10kw	20kw	30kw
射程	200m~800m	200m~2000m	200m~3000m	200m~4000m
射击角度	倾斜角度：-10°~80°			
射击方位	0°~360°			
截获距离	≧4km			
目标速度	≦60m/s			
目标尺寸	内径≦2m			
跟踪精度	高于15μ rad			
最大精准跟踪角速度	20°/s			
最大精准跟踪加速度	5°/s^2			
打击间隔	6s			
连续发射激光的最大时长	200s			
平均摧毁目标时间	10s			

"Silent Hunter" Anti-Terrorism Mission Vehicle. Consists of a C2 system and a radar and optical subsystem. Laser emitters target low-flying mortars and rockets. Poly Technologies. 2018 Defence Services Asia (Malaysia). 1/1.

TS-504
TROPOSCATTER COMMUNICATION VEHICULAR STATION

TS-504 Troposcatter Communication Vehicular Station. Poly Defense. 2016 Defense Services Asia (Malaysia). (1/2).

General Description

TS-504 Troposcatter Communication Vehicular Station is mainly used for point-to-point BLOS, long-range, multi-channel digital signal transmission. The services can be voice, fax and data (including IP data) etc.

Features

- Integrated design and good mobility
- With BLOS transmission capacity
- With octuple diversity receiving techniques; good anti-fading capability
- With BITE function; the fault will be located at modules
- The transmission rate can be 256/512/1,024/2,048 Kb/s
- Support voice, data and IP services
- With central monitor function
- With EOW function for inter-station communication
- With optical fiber connection function
- Data channel encryption

Specifications

Communication Range	
-Transmission Rate	2,048kb/s
-Distance (Dual Antenna)	150km
Operation Frequency	
-Model A	4,400MHz~4,580MHz (Tx Frequency)
	4,820MHz~5,000MHz (Rx Frequency)
-Model B	4,820MHz~5,000MHz (Tx Frequency)
	4,400MHz~4,580MHz (Rx Frequency)
-Stepping	1MHz
-Min Interval of Tx and Rx Frequency	≥240MHz
-Preset Channel No	12
Antenna	
-Antenna Equivalent Size	2.4m
-Polarization	Horizontal Linear Polarization
-Interface	BJ48 Waveguide
-Gain	≥33dBi

Functions

- Communication Service
- EOW
- Encryption
- Monitor
- Service Access

Add: 27/F, New Poly Plaza, No.1 Chaoyangmen Beidajie, Dongcheng District, Beijing, China
Post Code: 100010
Tel: (8610) 6408 2288   Fax: (8610) 6408 2988
www.poly.com.cn   E-mail: poly@polyinc.com

POLY DEFENCE

TS-504 Troposcatter Communication Vehicular Station. Poly Defense. 2016 Defense Services Asia (Malaysia). (2/2).

CHINA ELECTRONICS CORPORATION (CEC)

CEC 中国电子信息产业集团有限公司
CHINA ELECTRONICS CORPORATION

地址：北京市海淀区万寿路27号　邮编：100846
电话：010-68207015　传真：010-68213745
邮箱：chenjp@cec.com.cn　chenzhh@cec.com.cn
网址：www.cec.com.cn

Add: No.27 Wanshou Road, Beijing 100846, China
Tel:+8610-68207015　Fax:+8610-68213745
E-mail:chenjp@cec.com.cn　chenzhh@cec.com.cn
http://www.cec.com.cn

China Electronics Corp. (CEC). Booklet. 2016 Zhuhai Airshow. (1/54).

南京熊猫汉达科技有限公司
Nanjing Panda Handa Technology Co.,Ltd.

公司简介 Company Profile

南京熊猫汉达科技有限公司（简称"熊猫汉达"）是国内军用通信领域的标杆企业。主要从事卫星、短波、超短波等军用通信产品的研发、生产、销售和服务，产品广泛应用于陆、海、空三军及二炮、武警、国家安全等部门，是我国军用卫星、短波、超短波通信领域产品研发与生产的重要基地。

Nanjing Panda Handa Technology Co.LTD (Referring "Panda Handa") is a model enterprise in the field of domestic military communications. There are products of satellite- communication, shortwave, ultra shortwave and other military communications. It covers the whole lifecycle of product, such as researching, processing, sales and services. The production are widely used in every fields, such as the navy force, the air force, the ground force, the Second Artillery Corps, People's Armed Police, State Security Department and etc. Throughout all the military communication fields, it's the important base of researching and processing.

地址：南京市联合村3号 邮编:210014 电话:025-51802700 传真:025-84827489
Add: No. 3 Village Road, Qinhuai District, Nanjing, China   Zip code: 210014
Tel: +86 25-51802700   Fax: +86 25-84827489   E-mail: handa@panda.cn

China Electronics Corp. (CEC). Booklet. Nanjing Panda Handa Technology Co., Ltd. (1/4). 2016 Zhuhai Airshow. (2/54).

Ku频段直升机载站
The Ku Frequency Band Helicopter Station

该站型国内领先，创新性提出一体化射频方案，实现直升机动中通天线小型化与轻量化，信道采用标准ATR机箱，提出专用自适应抗旋翼遮挡算法，具备在旋翼遮挡条件下的Ku频段卫星通信能力，为直升机平台与地面指控中心之间提供超视距的话音、数据、视频双向传输功能。

The product is advanced in our country with the Ku frequency band's satellite communication capability under the condition of rotor occlusion. It has been designed with an innovative solution of integrated radio frequency, so that the antenna connecting in moving of helicopter is small in size and light in weight. The channel uses standard ATR cabinet, and it has been designed by using a special adaptive anti-rotor shielding algorithm. The product can provide transmission functions of the over horizon's voice、data and video between the helicopter platform and the ground control station.

尺寸：直径×高：φ600mm×420mm
Diameter×Height：φ600mm×420mm

China Electronics Corp. (CEC). Booklet. Nanjing Panda Handa Technology Co., Ltd. (2/2). The Ku Frequency Band Helicopter Station. 2016 Zhuhai Airshow. (3/54).

南京长江电子信息产业集团有限公司
Nanjing Changjiang Electronics Group Co.,Ltd.

公司简介 Company Profile

2016 zhuhai

南京长江电子信息产业集团有限公司是隶属于中国电子信息产业集团的国有大型军工电子企业。企业始建于1946年，是中国第一家设计、制造并出口大型电子装备的整机制造企业。公司先后研制生产了四十多种型号四千多部雷达装备，是中国对空防御、对海监视和舰载雷达的主要供应商之一。

南京长江电子信息产业集团占地五十多万平米，下属十余家子、分公司，拥有雷达研发中心、计量例试中心、数控加工中心、SMT加工中心和高科技产业园区。公司拥有先进的产品设计软件和检测手段，建有一套有效的质量控制体系，产品通过了GJB9001A-2001质量管理体系标准认证。主要雷达产品有远程对空防御三坐标雷达、远程对空警戒兼引导雷达、海岸警戒雷达、舰载搜索雷达和高机动低空警戒雷达及远程航路监视雷达等。

Nanjing Changjiang Electronics Group Co., Ltd.(CJEC) is a large state-owned company affiliated with China Electronics (CEC). Established in 1946, NJCJEC is the China's first radar manufacturer, and has provided more than four thousand sets of radar equipments for China airforce and navy with more than forty models. Nowadays, NJCJEC is one of the main suppliers of radar in China supplying air defence, sea surface targets surveillance and shipborne radar.

NJCJEC occupies area of more than 500 thousand square meters. It has ten subsidiary companies. It has over ten subsidiary companies. It owns a radar research and development center, a modern digital-controlled processing center, an advanced SMT processing center and a high-tech park. There are many advanced design software and testing measures. It has set up an effective quality control system. The military products have approved the GJB9001A-2001 quality management standard. Main radar products are long-range 3-D air defence radar, long-range air surveillance and guidance radar, coast surveillance radar, shipborne search radar, high mobile low air surveillance radar and long-range route surveillance radar etc.

地址：南京经济技术开发区恒谊路9号 邮编：210038 电话：025-85799387 传真：025-85800209
Add: 9 Heng Yi Road, Nanjing Economic and Technological Development Zone
Zip Code: 210038 Tel: +86-25-85799387 Fax: +86-25-85800209

China Electronics Corp. (CEC). Booklet. Nanjing Changjiang Electronics Group Co., Ltd. (NJCJEC). (1/2). 2016 Zhuhai Airshow. (4/54).

REL-4型远程三坐标雷达

REL-4型雷达是一部L波段、机动型、全相参脉冲压缩、机-相扫体制的三坐标雷达。在对空防御体系中担负目标引导和对空警戒任务，具有多种抗干扰手段，探测能力强，机动性好，能综合多部雷达信息完成数据融合及区域组网，具备远程遥控、在线性能测试能力。

REL-4 Long Range 3D Radar

REL-4 is an L-band, mobile type, full coherent pulse compression, mechanical-phase scan system 3D radar. It shoulders the task of target guidance and air surveillance in air defence system. The radar has multi-anti-jamming measures with strong detection capability and excellent mobility. It can integrate and manage information from several radars to complete data fusion and forming area network. It has the ability of long range remote control and on line performance test.

REL-2A型岸用对海对低空警戒雷达

REL-2A型雷达为L波段、全固态、脉冲压缩体制的对海、对低空警戒雷达。适用于海岸对海、对低空监视，具有可靠性好、抗干扰能力强等优点。可为岸防武器系统提供目标指示，为舰船实施辅助导航，能与敌我识别器协同工作，完成敌我识别。

REL-2A Coast Sea and Low Air Surveillance Radar

REL-2A is a L-band full solid-state coast sea and low air surveillance radar with pulse compression technique features. It is used for sea and low air surveillance with good reliability and strong anti-jamming capability. It can provide target indication for coastal defense arm system and to make assistant navigation for our vessels and warships. The radar cooperates with IFF interrogator to complete character identifying.

REC-1/1S型舰载搜索雷达

REC-1/1S型雷达是C波段、全相参、脉冲压缩体制的舰用对海对空搜索雷达，用于中小型水面舰船作目标指示，为本舰的舰炮武器系统提供目标参数。具有独立的海、空处理通道，优良的低空探测能力和抗干扰性能，是在复杂环境下完成使命任务的高性价比舰用雷达装备。

REC-1/1S Shipborne Search Radar

REC-1/1S is a C-band shipborne sea and air search radar with full coherent and pulse compression technique features. It is used as target indication for small and medium size ships and provide target data for ship cannon arm system itself. The radar has independent sea and air processing channels with excellent low air detection capability and anti-jamming performance and high rate of performance and price. It can complete mission and task under complicated environment.

China Electronics Corp. (CEC). Booklet. Nanjing Changjiang Electronics Group Co., Ltd. (NJCJEC). (2/2). REL-4 Long Range 3D Radar; REL-2A Coast Sea and Low Air Surveillance Radar; REC-1/1S Shipborne Search Radar. 2016 Zhuhai Airshow. (5/54).

CEC 中国电子

南京科瑞达电子装备有限责任公司
NANJING CORAD ELECTRONIC EQUIPMENT CO., LTD.

公司简介 Company Profile

南京科瑞达电子装备有限责任公司始建于20世纪50年代，是中国最早的从事特种电子装备研制生产专业厂家之一，现隶属于中国电子信息产业集团公司（CEC）。公司现主要从事机载、舰载、便携式雷达侦察干扰设备等电子对抗装备的研制，是中国雷达对抗装备的主要供应商。

Nanjing Corad Electronic Equipment Co., Ltd. was founded in 1950s. It is treated as one of the pioneer professional factories of China focused on special electronic warfare equipment, which is subjected to CEC now. It develops and product the airborne, ship-borne, portable EW equipments, which is one of the main EW equipment suppliers for PLA.

地址：南京市江宁经济技术开发区天元西路111号
邮编：211100 电话：025-68805801 传真：025-68805858
Add: No.111 Tian Yuan Xi Lu, Jiang Ning Economic District, Nanjing, China
Postcode: 211100 Tel: +86 025-68805801 Fax: +86 25-68805858

China Electronics Corp. (CEC). Booklet. Nanjing Corad Electronic Equipment Co., Ltd. (1/4). 2016 Zhuhai Airshow. (6/54).

ERR107A便携式雷达侦察设备
ERR107A Portable Radar Reconnaissance Equipment

ERR107A便携式雷达侦察设备主要用于检测战场监视雷达、炮位侦察雷达,适用于前沿阵地、高原、山地的单兵作战,也可以供边防哨所使用。

ERR107A Portable Radar Reconnaissance Equipment is mainly designed to detect and receive signals from battlefield surveillance radars, artillery locating radars. It is suitable for individual military operator to fight in forward position, highlands and mountainous regions, and for border sentry posts as well.

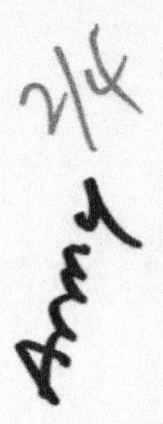

地址:南京市江宁经济技术开发区天元西路111号
邮编: 211100 电话: 025-68805801 传真: 025-68805858
Add: No.111 Tian Yuan Xi Lu, Jiang Ning Economic District, Nanjing, China
Postcode: 211100 Tel: +86 025-68805801 Fax: +86 25-68805858

China Electronics Corp. (CEC). Booklet. Nanjing Corad Electronic Equipment Co., Ltd. (2/4). ERR107A Portable Radar Reconnaissance Equipment. 2016 Zhuhai Airshow. (7/54).

SRW210A雷达侦察告警设备
SRW210A Radar Reconnaissance Equipment

SRW210A是先进的舰载ESM设备，用于侦收2~40GHz频率范围内各种体制的雷达信号，进行参数测量，并自动进行分析，威胁判断和威胁告警，并具有无源干扰资源管理和战术决功能，适装于导弹快艇、护卫舰、驱逐舰等水面作战舰艇。

SRW210A Radar Reconnaissance Equipment is an advanced shipborne ESM equipment, intercept and collect radar signals on the frequency range from 2 to 40 GHz, measure the radar operating parameters, automatical analysis, threat evaluation and threat alarm, manage the passive jamming resources and make tactic decision, is suitable for various types of combat ships, such as boats, frigates and destroyers.

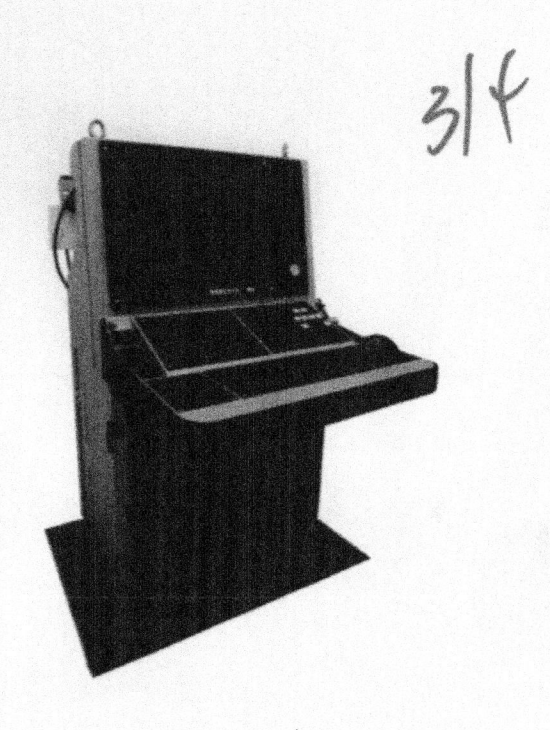

地址：南京市江宁经济技术开发区天元西路111号
邮编：211100 电话：025-68805801 传真：025-68805858
Add: No.111 Tian Yuan Xi Lu, Jiang Ning Economic District, Nanjing, China
Postcode: 211100 Tel: +86 025-68805801 Fax: +86 25-68805858

China Electronics Corp. (CEC). Booklet. Nanjing Corad Electronic Equipment Co., Ltd. (3/4). SRW210A Radar Reconnaissance Equipment. 2016 Zhuhai Airshow. (8/54).

ARW9101机载雷达告警设备
ARW9101 Airborne Radar Warning Receiver

ARW9101是先进的机载全向雷达告警设备，主要用来侦收、分析、识别来各种体制的雷达信号，测定雷达的主要参数和方位，确定雷达的工作方式，工作状态，自动判别目标性质和威胁程度，实时向飞行员发出威胁告警。

ARW9101 Airborne Radar Warning Receiver is an advanced airborne omnidirectional radar warning equipment, is designed to receive, analyze, identify radar signals from various radars, measure radars' parameters and bearing, determine radars' operation mode and operation state, automatically distinguishes targets' nature and degree of threat, provides a real-time alarm to the pilot.

地址：南京市江宁经济技术开发区天元西路111号
邮编：211100 电话：025-68805801 传真：025-68805858
Add: No.111 Tian Yuan Xi Lu, Jiang Ning Economic District, Nanjing, China
Postcode: 211100 Tel: +86 025-68805801 Fax: +86 25-68805858

China Electronics Corp. (CEC). Booklet. Nanjing Corad Electronic Equipment Co., Ltd. (4/4). ARW9101 Airborne Radar Warning Receiver. 2016 Zhuhai Airshow. (9/54).

武汉中原电子集团有限公司
Wuhan Zhongyuan Electronics Group Co., Ltd.

公司简介 Company Profile

武汉中原电子集团有限公司位于武汉东湖新技术开发区（中国光谷），是研制生产无线通信设备、电子系统工程设备、电子应用产品及各类电池的高新技术企业，是国防战术通信装备研制、生产及服务的重点骨干企业。公司拥有4个研发机构、1个国防计量站/军用实验室和1个博士后科研工作站，具有通过认证持续运行有效的质量管理体系，具有武器装备科研生产单位二级保密资格，是总装备部装备承制单位，并持有武器装备科研生产许可证。

Zhongyuan Electronics Group Co.,Ltd. is located at DongHu New technology development district of WuHan(Optical Valley of China).It is a hi-tech enterprise group researching in wireless communication equipment、electronics system engineering equipment、electronics application products and various of batteries, also it is the backbone enterprise of national defense tactical communication equipment research、manufacture and service.

The company has four research and development institutions、one national defense metrology station/military lab and one Doctor scientific research station. The company passed ISO9001 quality identification and ISO10012.1 measurement identification. The product takes self-supporting export rights. It is the equipment contractor of GAD, also it owns the certificate of weapon equipment research and manufacture production right.t's the important base of researching and processing.

地址：武汉市东湖新技术开发区高新四路1号 邮编：430205 电话：027-81789710 传真：027-81788710
Add: GaoXin 4th road No.1, DongHu New technology development district, WuHan, China
Zip code: 430205 Tel: +86 27-81789710 Fax: +86 27-81788710 E-mail: zhongyuan710@yahoo.com.cn

China Electronics Corp. (CEC). Booklet. Wuhan Zhongyuan Electronics Group Co., Ltd. (1/4). 2016 Zhuhai Airshow. (10/54).

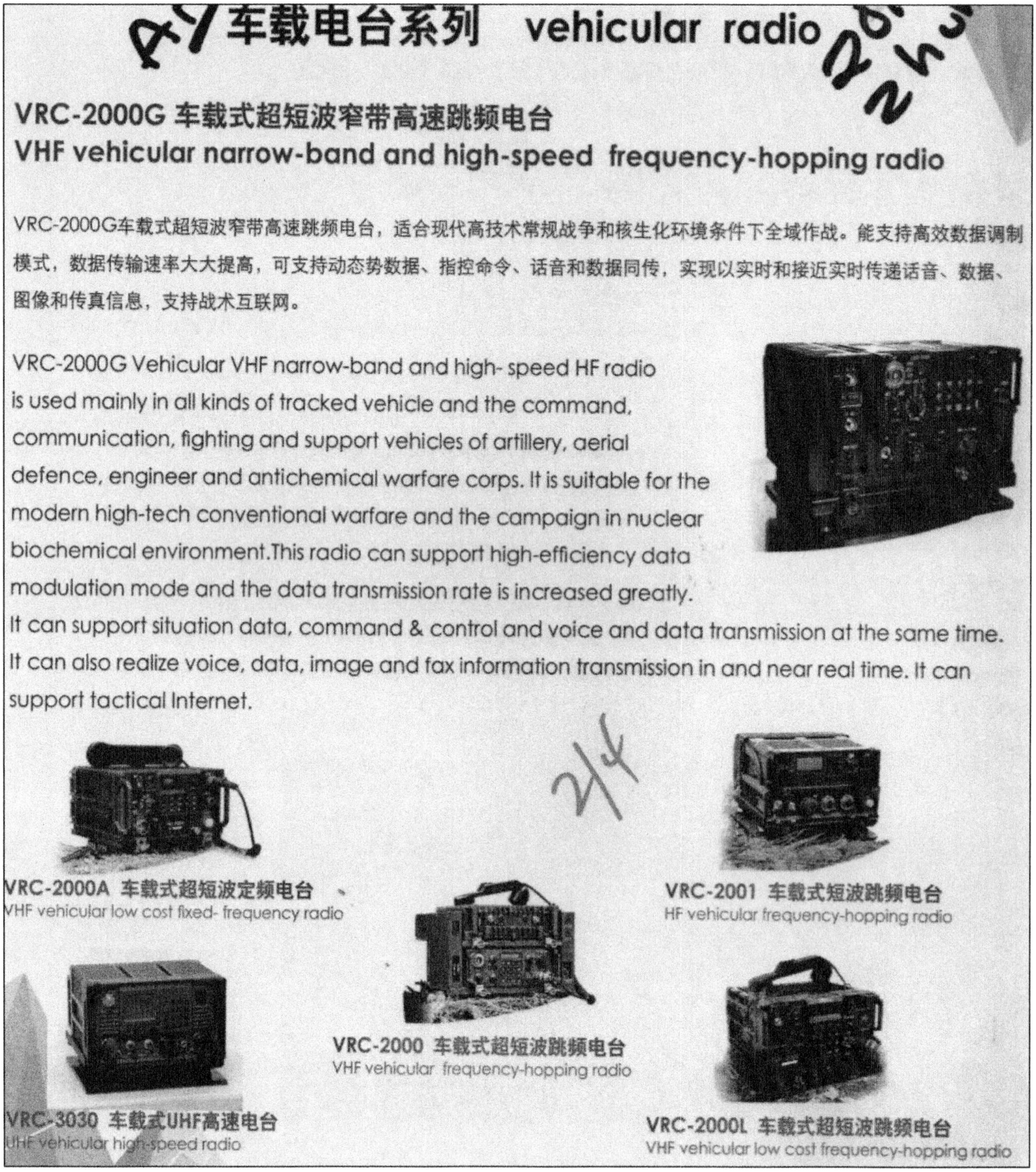

车载电台系列 vehicular radio

VRC-2000G 车载式超短波窄带高速跳频电台
VHF vehicular narrow-band and high-speed frequency-hopping radio

VRC-2000G车载式超短波窄带高速跳频电台，适合现代高技术常规战争和核生化环境条件下全域作战。能支持高效数据调制模式，数据传输速率大大提高，可支持动态势数据、指控命令、话音和数据同传，实现以实时和接近实时传递话音、数据、图像和传真信息，支持战术互联网。

VRC-2000G Vehicular VHF narrow-band and high-speed HF radio is used mainly in all kinds of tracked vehicle and the command, communication, fighting and support vehicles of artillery, aerial defence, engineer and antichemical warfare corps. It is suitable for the modern high-tech conventional warfare and the campaign in nuclear biochemical environment. This radio can support high-efficiency data modulation mode and the data transmission rate is increased greatly. It can support situation data, command & control and voice and data transmission at the same time. It can also realize voice, data, image and fax information transmission in and near real time. It can support tactical Internet.

VRC-2000A 车载式超短波定频电台
VHF vehicular low cost fixed-frequency radio

VRC-2001 车载式短波跳频电台
HF vehicular frequency-hopping radio

VRC-2000 车载式超短波跳频电台
VHF vehicular frequency-hopping radio

VRC-3030 车载式UHF高速电台
UHF vehicular high-speed radio

VRC-2000L 车载式超短波跳频电台
VHF vehicular low cost frequency-hopping radio

China Electronics Corp. (CEC). Booklet. Wuhan Zhongyuan Electronics Group Co., Ltd. (2/4). Vehicular Radio: VRC-2000, VRC-2000A; VRC-2000L; VRC-2001; VRC-3030. 2016 Zhuhai Airshow. (11/54).

背负电台系列 manpack radio

PRC-2000G 背负式超短波窄带高速跳频电台
VHF manpack narrow-band and high-speed frequency-hopping radio

PRC-2000G背负式超短波窄带高速跳频电台，主要运用于单兵作战系统。能支持高效数据调制模式，数据传输速率大大提高，可支持动态势数据、指控命令、话音和数据同传，实现以实时和接近实时传递话音、数据、图像和传真信息，支持战术互联网，为陆军机械化部队实施作战指挥控制提供了高速数据通信链路，满足集团军信息系统集成和应急作战部队信息化建设的需要。

PRC-2000G VHF manpack narrow-band and high-speed FH radio series is used mainly in single soldier fighting system. With the high-efficiency data modulation mode, the data transmission rate of this radio is enhanced greatly. It can support situation data, command & control and voice and data transmission at the same time. It can also realize voice, data, image and fax information transmission in and near real time. It can support the tactical Internet, which provides the high-speed data communication link to command and control for land army. So it meets the information system integration requirements for a group army and the informatization construction requirements of the emergency fighting army.

PRC-2001 背负式短波跳频电台
HF manpack frequency-hopping radio

TBR-2000 背负式超短波跳频电台
VHF manpack frequency-hopping radio

China Electronics Corp. (CEC). Booklet. Wuhan Zhongyuan Electronics Group Co., Ltd. (3/4). PRC-2000G VHF Manpack Narrow-band and High-band Frequency-Hopping Radio. 2016 Zhuhai Airshow. (12/54).

手持电台系列 handheld radio

PRC-2000H 手持式超短波跳频电台
VHF handheld frequency-hopping radio

PRC-2000H手持式超短波跳频电台主要运用于班排通信或单兵作战系统。具有语音、数据功能，为陆军单兵作战系统提供了数据通信链路，满足集团军信息系统集成和应急作战部队信息化建设的需要。电台为通用战术电台，可与VRC-2000系列电台兼容互通。

PRC-2000H Handheld VHF frequency-hopping radio is used mainly for squad/platoon communication and single soldier fighting system. It has voice, data, which has provided data communication link to command & control for land army. So it meets the information system integration requirements of a group army and the informationization construction requirements of emergency fighting army.

PRC-2000S 手持式超短波定频电台
VHF handheld frequency-fixed radio

PRC-2000D 手持式超短波定频电台
VHF handheld frequency-fixed radio

VIC-2001H 数字型车内通话器
VIC-2001H Vehicle Digital Intercom System

PVIC-2001H数字车内通话系统采用数字话音技术和RS－485异步半双工总线传输方式，总线速率高达1536K，话音采用64K语音编解码方式。该系统可配接VRC-2000系列电台。可供坦克、装甲车、炮车等车辆内的乘员作车内通话或控制电台对外进行通信、转信等。

VIC-2001H As the intercom adapts the digital voice technology and RS–485 asynchronous semi-duplex bus transmission mode, the bus rate is up to 1536k, and the voice uses 64k voice coding and decoding modes. he system is used for crew members in tanks, armored vehicles, artillery prime mover, etc to intercommunicate or control the radio to communicate, retransmit, etc.

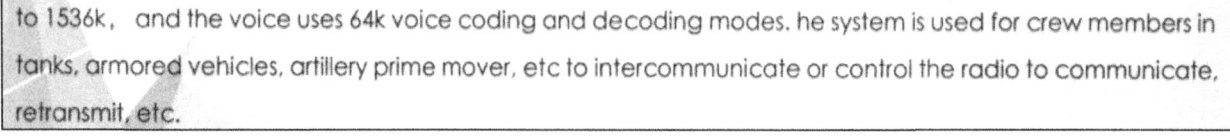

China Electronics Corp. (CEC). Booklet. Wuhan Zhongyuan Electronics Group Co., Ltd. (4/4). Handheld Radio: VIC-2000H Handheld Frequency-Hopping Radio; VIC-2001H Vehicle Digital Intercom System. 2016 Zhuhai Airshow. (13/54).

桂林长海发展有限责任公司
Guilin Changhai Development Co., Ltd.

Comany Profile

桂林长海发展有限责任公司（国营第七二二厂），隶属于中国电子信息产业集团有限公司。主要从事雷达、电子对抗、智能安防等领域的高新电子装备研制生产。公司位于桂林芦笛岩景区桃花江畔，占地面积80万平方米，具有从事国防工业的完整资质和科研生产条件。拥有一批国家级型研和预研项目，是国有大型整机高新电子企业。

Guilin Changhai Development Co., Ltd. (State-run No.722 Factory) is subordinate to the China Electronics Corporation (CEC), mainly concentrates on researching & manufacturing of radar, ECM, intelligent security, and other high-tech electronics. The company is situated besides the Taohua River and the Ludi Cave Scenic Area, covers an area of 800,000 square meters, and possesses the complete qualifications and researching & manufacturing conditions of being engaged in the national defense industry. The company also possesses a batch of national type approval research/beforehand research projects, is one of the large state-run high-tech electronic enterprises.

地址：广西桂林市长海路3号 邮编：541001 电话：0773-2691746 传真：0773-2694030
Add: No.3 Changhai Road, Guilin City, Guangxi, China Zip:541001
Tel: +86-773-2691746 Fax:+86-773-2694030

China Electronics Corp. (CEC). Booklet. Guilin Changhai Development Co., Ltd. (State-run No. 722 Factory). (1/4). 2016 Zhuhai Airshow. (14/54).

摆放式合成孔径雷达对抗设备

摆放式合成孔径雷达对抗设备可部署于我演习区域、沿海岛礁、导弹靶场和军事工程施工场地，对我军战术布局、地形等战场要素进行电子干扰防护。可降低敌方侦察机的探测精度，减弱其目标跟踪能力，并能减小监视覆盖区域，进而可以模糊其瞄准和攻击决策链。此外，设备部署于远海区域阵地，作为前沿侦察传感终端，可对工作频率范围内的雷达信号进行采集、分选和存储，为我军电子战情报收集等提供技术参数支撑。

Transportable SAR Jammer

Transportable SAR Jammer can be deployed in friendly exercise area, coastal reef, missile range and military engineering construction site, implementing electronic jamming protection for friendly tactical layout, terrain and other battlefield elements. The Jamming can reduce the detection accuracy of the hostile reconnaissance aircraft, weaken the ability of target tracking, and also can reduce the monitoring coverage area, thus it can fuzz hostile aiming/attack decision-making chain. Furthermore, the jammer can be deployed in pelagic region front, as a frontier reconnaissance sensor terminal, it can implement operations of capture, sort and storage for the radar signal within the operational frequency range, and provide technical support for our military electronic warfare intelligence collection.

China Electronics Corp. (CEC). Booklet. Guilin Changhai Development Co., Ltd. (State-run No. 722 Factory). (2/4). Transportable Synthetic Aperture Radar (SAR) Jammer. 2016 Zhuhai Airshow. (15/54).

便携式通信干扰设备

便携式通信干扰设备用于战时干扰敌方无线定频、跳频电台，阻断敌方通信，使敌人不能及时协调作战，掌握战场态势；也可用于非战时部队演练，提高部队适应复杂电磁环境的电子对抗能力

Portable Communication Jammer

Portable Communication Jammer is utilized in wartime to interfere with the hostile constant frequency/frequency hopping radio station, intercept hostile communication, in order to ensure that the hostile cannot obtain the battlefield situation, and cannot be coordinated in a timely manner. Moreover, the jammer can also be utilized in troop training in peacetime to enhance troop ECM capability in complex electromagnetic environment.

China Electronics Corp. (CEC). Booklet. Guilin Changhai Development Co., Ltd. (State-run No. 722 Factory). (3/4). Portable Communication Jammer. 2016 Zhuhai Airshow. (16/54).

便携式战场监视雷达

该雷达可用于地面，湖泊以及海面等具有战略意义目标和区域的防护，如反恐、防海盗等；同时也可用于民用设施的防护，如输油管道、钻井平台、民用机场、核电站和其它特殊设施和场所的保卫和监控

Portable Battlefield Surveillance Radar

The radar can be utilized for anti-piracy, anti-terrorism and other missions in strategically significant targets and regions, including ground, lake and sea surface. Moreover, the radar also can provide protection and monitoring for civilian targets, including oil pipeline, drilling platform, civil airport, nuclear power plant and other special facilities and places.

China Electronics Corp. (CEC). Booklet. Guilin Changhai Development Co., Ltd. (State-run No. 722 Factory). (4/4). Portable Battlefield Surveillance Radar. 2016 Zhuhai Airshow. (17/54).

北京圣非凡电子系统技术开发有限公司
BEIJING SHENGFEIFAN ELECTRONIC SYSTEM TECHNOLOGY DEVELOPMENT CO., LTD.

公司简介 Company Profile

2016 zhuhai

北京圣非凡电子系统技术开发有限公司隶属于中国电子信息产业集团，从1986年开始通信产品研制生产，至今已有四十余种型号的通信产品列装。在甚低频和高频（短波）等通信领域居国际先进、国内领先的地位。

BEIJING SHENGFEIFAN ELECTRONIC SYSTEM TECHNOLOGY DEVELOPMENT CO., LTD. is affiliated to CHINA ELECTRONIC CORPORATION (CEC). It started developing and manufacturing communication products since 1986. Since then, SHENGFEIFAN has developed more than 40 types of communication products that have been widely equipped in troops. Now, it has became the leading provider in Very Low Frequency(VLF), High Frequency (Short Wave) communication products industry.

1/2

北京圣非凡电子系统技术开发有限公司
BEIJING SHENGFEIFAN ELECTRONIC SYSTEM TECHNOLOGY DEVELOPMENT CO.,LTD

地址：北京市昌平区未来科技城南区蓬莱苑南路中国电子网络安全与信息产业基地B栋
邮编：102209　电话：010-66608124　传真：010-66608100
Add: Building B, China Electronics Cyber Security Informationization Industrial Base, Southearn Area of Weilaikeji-cheng, ChangPing, Beijing.　Zip code: 102209
Tel: +86 01-66608124　Fax: +86 01-66608100　E-mail: sff@cec-cesec.com.cn

China Electronics Corp. (CEC). Booklet. Beijing Shengfeifan Electronic System Technology Development Co., Ltd. (1/2). 2016 Zhuhai Airshow. (18/54).

车载机动甚低频发信系统
Introduction of Vehicle mobile Very Low Frequency (VLF) Transmission

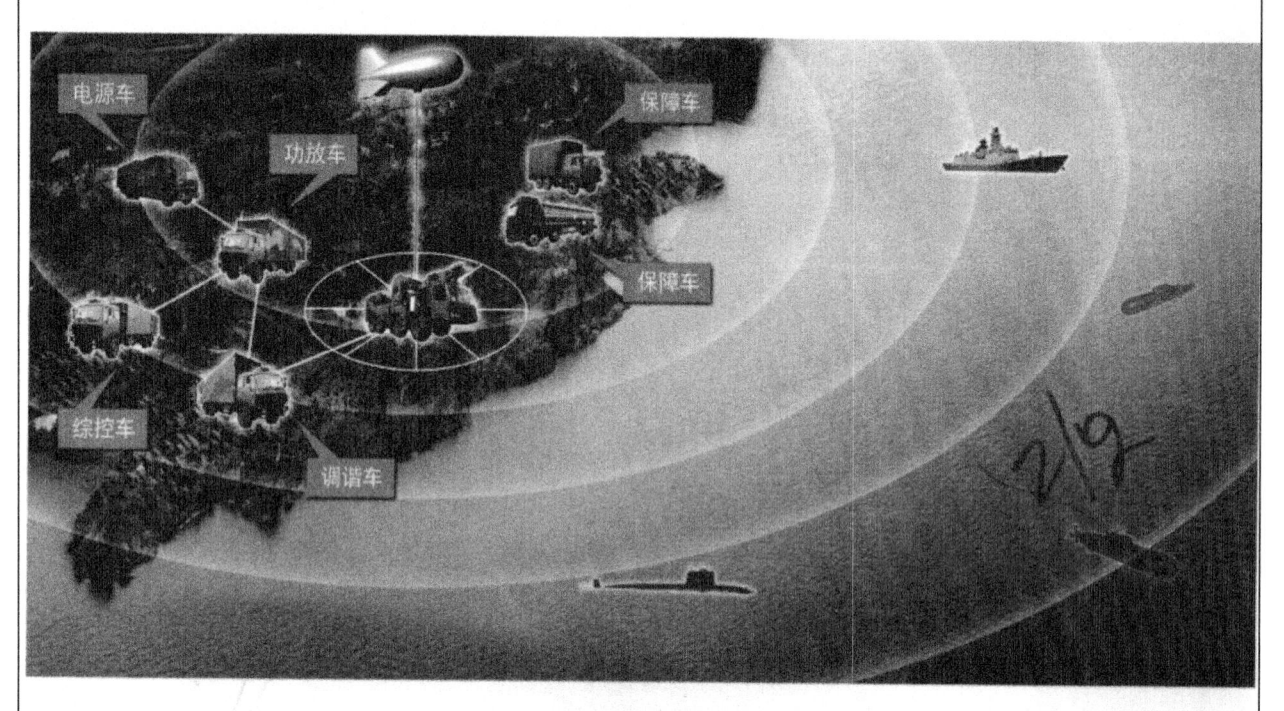

车载机动甚低频发信系统由甚低频功放车、甚低频调谐车、气球天线锚泊车、综合控制车、电源车及辅助车辆组成，系统可在不依赖其它设施的条件下，独立完成甚低频发信任务。车载机动甚低频发信系统全部设备均安装在轮式车辆上，机动性强、展开/撤收时间短、操作简便、可靠性高，并具有一定的抗毁能力。

Vehicle Mobile Very Low Frequency (VLF) Transmission System is composed of VLF Power Amplifier Vehicle, VLF TuningVehicle, Balloon Antenna AnchorVehicle, Comprehensive Control Vehicle, Power Supply Vehicleand AuxiliaryVehicle. The System can independently complete the mission of VLF transmission without other facilities involved.Vehicle Mobile Very Low Frequency (VLF) Transmission System are all equipped on wheeled vehicle, which has a lot of advantages in the aspect of mobility,flexibility,operability, reliability and survivability.

China Electronics Corp. (CEC). Booklet. Beijing Shengfeifan Electronic System Technology Development Co., Ltd. (2/2). Vehicle Mobile Very Low Frequency (VLF) Transmission System. 2016 Zhuhai Airshow. (19/54).

成都中电锦江信息产业有限公司
CEC-Jinjiang info industrial Co., Ltd.

公司简介 Company Profile

成都中电锦江信息产业有限公司（国营第七八四厂），位于素有"天府之国"之称的四川成都，1958年建成投产，是国家"一五"期间156项重点建设项目之一。公司是西部地区唯一的专业从事军民用地面雷达等电子系统工程产品研发、生产、经营的电子信息行业高科技企业，是国内地面雷达行业综合实力排名前列的军工电子骨干企业，是我国军民用新一代天气雷达的主要供应商，也是我国首批机电产品出口基地企业。

CEC-Jinjiang info industrial Co., Ltd, built in 1958 and located in Chengdu (called as "Land of Abundance"), Sichuan Province, China, is one of the 156 key projects built in the first national 5-year Plan period. It is the only high-tech enterprise in the electronic industry in western China area for research, development, manufacture and marketing of electronic system products of military and civil ground radars. It ranks in the forefront as a backbone military electronic enterprise in ground radar sector in China, and is also a major supplier of military and civil new generation weather radars. It was selected as one of the first export base enterprises for electric machinery products.

地址：中国成都市建设北路三段168号　邮编：610051　电话：028-84394882
传真：028-84394281　邮箱：SCS784@163.COM　网页：www.jec784.com
Add: No.168, Section 3, Jian She Bei Road, Chengdu, China
Zip code: 610051　Tel: 028-84394882　fax: 028-84394281
E-mail: SCS784@163.COM　Web: www.jec784.com

China Electronics Corp. (CEC). Booklet. Jinjiang Info Industrial Co., Ltd. (1/3). 2016 Zhuhai Airshow. (20/54).

China Electronics Corp. (CEC). Booklet. Jinjiang Info Industrial Co., Ltd. (2/3). 2016 Zhuhai Airshow. Meteorological and Hydrological Products: 714DN-D Dual Backup Weather Radar; 714XDP-M Weather Radar; PR-11A First Precipitation Monitoring Radar; 714CDP-K Weather Radar; X-band Portable Dual Polarization Weather Radar System (21/54).

JL3D-91B 米波三坐标雷达简介
Introduction to JL3D-91B Radar

JL3D-91B米波三坐标雷达（以下简称91B雷达）是新一代的米波三坐标中远程对空情报雷达，具有反隐身性能优越、探测威力大、测量精度高、自动化程度高、可靠性好、抗干扰能力强等特点，并具有良好的机动性。

该雷达主要面向中远程警戒雷达的国际中高端市场，它可以满足现代防空系统中警戒探测的多频段搭配部署和反隐身目标的需求，能以较高性价比满足国土防空需要。

JL3D-91B metric wave 3D radar (referred as 91B hereafter) is a new generation metric wave medium/far range air surveillance radar featuring predominant anti-stealth performance, powerful detection coverage, high automatization, good reliability and outstanding anti-jamming capability as well as excellent maneuverability.

This radar is designed for middle/high-end market in medium/far range surveillance radar. It can meet the requirements both on anti-stealth target detection and multi-band combined deployment in modern air defense system; furthermore, its higher cost performance can well satisfy the demand of the national air defense in many countries.

主要技术特点
- 隐身目标的探测优势
- 高速高机动目标的跟踪能力强
- 探测空域大
- 高精度测量
- 战场适应能力强
- 情报综合、上报能力
- 良好的可靠性和维修性
- 高机动性部署
- 环境适应能力强

Main Technical Features
- Innate Advantage on Stealth Targets Detection
- Capability of Tacking High-Speed/Maneuverability Targets
- Large Detection Coverage
- High Accuracy Measurement
- Strong Battle Field Adaptability
- Intelligence Synthesis and Report
- Good Reliability and Maintainability
- High Maneuverability
- Excellent Environmental Adaptability

China Electronics Corp. (CEC). Booklet. Jinjiang Info Industrial Co., Ltd. (3/3). 2016 Zhuhai Airshow. JL3D-91B Radar. See also: Minnick, Wendell, *Chinese Radars: Product Brochures* (Amazon 2017). (22/54).

南京三乐电子信息产业集团有限公司
Nanjing Sanle Electronics Gorporation

公司简介 Company Profile

南京三乐电子信息产业集团有限公司直属中国电子，是中国真空电子行业的摇篮，研制出中国第一只电子管。

公司真空电子和光电子产品应用于雷达、电子对抗、情报侦察、通讯导航、载人航天等领域；微波装备应用于生物质热解、金属冶炼、煤炭处理等领域。公司将围绕市场需求持续创新，为客户创造良好的社会效益和经济效益。

Nanjing Sanle Electronics Gorporation is regarded as the cradle of Vacuum Electronic Industry in China. It's predecessor is the National Research Laboratory established in 1935, and was expanded to the first pecialized Electrin Tube Work in 1951.

We have strong ability in R & D and technology in vacuum electronics. Our main products include vacuum electronic elements, microwave application system, special glass and ceramics, electrical vacuum equipments, etc. Since our company was founded, we have developed more than 600 series of new products & technology. Among them, nearly 100 series fill in the gap in domestic industries, and more than 300 series won all kinds of awards from the nation.

Technology achieves success, intelligence leads to future. During the coming years, we will accelerate to adjust our products & industry structure, commit to build our industry chain and devote to become the most advanced R & D base in microwave vacuum electronic elements and application equipments all over the world.

地址：南京市浦口区光明路5号 邮编：211800 电话：025-83731417 传真：025-83733213
Add: No.5, Guangminglu, Pukou, Nanjing, China
Postcode: 211800 Tel: 025-83731417 Fax: 025-83733213 E-mail: 360734641@qq.com

China Electronics Corp. (CEC). Booklet. Nanjing Sanle Electronics Corp. (1/4). 2016 Zhuhai Airshow. (23/54).

幅相一致性脉冲行波管 BM-2021F
Phase-amplitude-matched pulsed TWT BM-2021

BM-2021F 管是 C 波段栅控幅相一致性脉冲行波管，该管总体结构采用螺旋线慢波系统、钐钴永磁包装、金属陶瓷结构，底板风冷结构。该产品用于高功率微波放大器系统，每套系统由多只幅相一致行波管组阵合成，用作微波功率放大。

BM-2021F is a kind of grid-controlled phase-amplitude-matched pulsed TWT working on C-band. It consists of helix slow-wave structure, Sm-Co electro-magnetic focusing system, ceramic materials, metals and air-cooling structure. BM-2021F is used in high power microwave amplifier system. Each system contains a group of several phase-amplitude-matched TWTs.

主要性能指标：
Technical Data:

工作频率： / Frequency:	C 波段 / C-band
热丝快启动电压： / Quick-start filament voltage:	5.5V ± 1V
脉冲输出功率： / Pulsed output power:	≥ 2.2kW
工作比： / D(Duty Cycle):	4.5% (Max)
相位一致性： / Phase consistency:	≤ |±25°|
幅度一致性： / Amplitude consistency:	≤ |±1dB| (饱和输出时) / ≤ |±1dB| (with saturated output)
外形尺寸： / Overall dimension:	≤ 500mm × 137mm × 112mm
重量： / Weight:	≤ 4.5 kg
输能方式： / RF input and output method:	SMA-F 输入、N 型-F 输出 / SMA-K coaxial input, N-F coaxial output

China Electronics Corp. (CEC). Booklet. Nanjing Sanle Electronics Corp. (2/4). Phase-Amplitude-Matched Pulsed Traveling Wave Tubes (TWT) BM-2021F. 2016 Zhuhai Airshow. (24/54).

TC-527A 型光电探测器
TC-527A photo-electric detector

一、结构特点和用途 (Structure characteristics and Application)

该探测器采用直径为 25mm 端窗式半透明锑钾铯光电阴极，光谱响应范围为 (300~650) nm，峰值波长为 (400±20) nm，并采用 13 级百叶窗式倍增系统，该探测器具有抗振性能好，坪区范围宽等特点，由特殊材料灌封成一体化，抗冲击振动；主要用于航天器和车载武器系统等领域。

The structure of the The TC-527A is the 25mm diameter head-on 13-stage photo-electric detector having a semitransparent bialkali photocathode and a venetian bline type multiplier system. The TC-527A has a spectral response characteristic at the range of 300~650 nm with the wavelength peak located at 400±20 nm. And the detector pots into integration by employing unique materials to defend impact vibration condition. Due to It's remarkable anti-seismic performance and broadly plateau characteristic, the detector is primary used in space vehicle and vehicular weapon system etc.

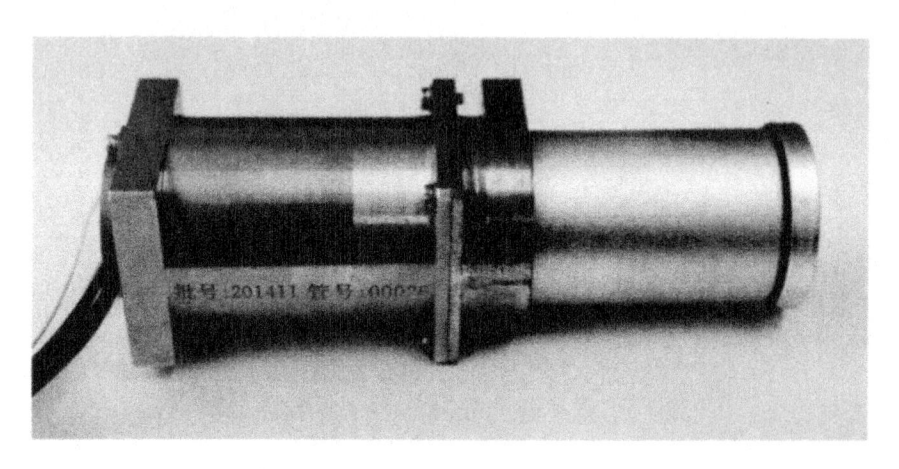

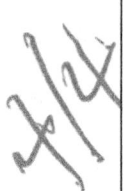

二、主要指标 (Technical parameter)

阴极特性 (Photocathode Parameters): ≤2100V;
紫外灵敏度 (Itraviolet Radiant Sensitivity): ≥12 mA/W°;
对 137Cs 幅度分辨率 (Pulse Amplitude Distribution Curve of The Radioactive Source 137Cs): ≤18%
坪长 (Plateau Length): ≥150V;
坪斜 (Plateau slope): ≤8%/100V;
额定寿命 (Life): 500h.

China Electronics Corp. (CEC). Booklet. Nanjing Sanle Electronics Corp. (3/4). TC-527A Photo-Electronic Detector. 2016 Zhuhai Airshow. (25/54).

L 频段空间行波管
L-band space TWT

L 频段空间行波管总体结构采用阳控电子枪、螺旋线慢波系统、四级降压收集极、钐钴永磁包装、金属陶瓷结构。该管在北斗导航卫星上使用。

L-band space TWT consists of anode-controlled electron gun, helix slow-wave structure, four-stage depressing collector, Sm-Co electro-magnetic focusing system, ceramic materials and metals. It is used in orbiter of the Beidou system.

主要性能指标：
Technical Data:

工作频率 / Frequency:	L 波段 / L-band
阳极电压 / Anode voltage:	3.8±0.2kV
输出功率 / Output power:	157~162W
工作方式 / Operation mode:	连续波 / continuous wave
寿命 / Lifespan:	≥12年 / ≥12 years
重量 / Weight:	≤2.55kg
外形尺寸 / Overall dimension:	≤585mm×85mm×73mm
输能方式 / RF input and output method:	SMA 同轴输入 / SMA coaxial input; TNC 同轴输出 / TNC coaxial output

China Electronics Corp. (CEC). Booklet. Nanjing Sanle Electronics Corp. (4/4). L-band space TWT. 2016 Zhuhai Airshow. (26/54).

南京华东电子集团有限公司
Nanjing Huadong Electronics Group Co., Ltd.

公司简介 Company Profile

2016 Zhuhai

南京华东电子集团有限公司是一家具有70多年悠久历史国有军工企业，是中国CRT行业中多个品种第一只产品的诞生地及发源地。主要从事液晶显示器件、触摸屏、晶体元件、导光板、被复线宽带综合数字远传通信设备、安澜可信操作系统、军人心理智能测评及训练系统、智能手环、北斗定位监控器、飞行膝板、地图作业终端、平板电脑、汉凡交互式电子手册等20多个门类、300多个品种产品的研发、生产、销售及服务，产品应用范围涵盖国民经济和国防事业的各个领域，同时还远销美国、德国、意大利、日本、罗马、越南、马来西亚、韩国、印度等国家和地区。

Nanjing Huadong Electronics Group Co., Ltd. is a long history of 70 years of state-owned military enterprises, is China's CRT industry in a number of varieties of the birth place of the first product and birthplace. Mainly engaged in liquid crystal display device, touch screen, crystal element, light guide plate, by double broadband integrated digital remote communication equipment, Alan trusted operating system, psychological intelligence assessment and training system, smart bracelet, Beidou Positioning monitor, flying knee plate, map operation terminal and tablet computer, Hanfan IETM and other more than and 20 categories, more than and 300 varieties of product development, production, sales and service, product application covers all areas of the national economy and national defense, but also exported to the United States, Germany, Italy, Japan, Rome, Vietnam, Malaysia, South Korea, India and other countries and regions.

地址：中国江苏省南京市华电路1号 邮编：210028 电话：025-52686115 传真：025-85319859
Add: 1# Hua Dian Road, NanJing, Jiangsu Province, China   Zip code:210028
Tel: 025-52686115   Fax: 025-85319859   E-mail: 19876074@163.com

China Electronics Corp. (CEC). Booklet. Nanjing Huadong Electronics Group Corp., Ltd. (1/4). 2016 Zhuhai Airshow. (27/54).

产品介绍 Product Introduction

NO.1 加固液晶显示器 / Reinforced liquid crystal display
机载、舰载、车载等显控设备
Redundancy and safety assurance for any safety-critical applications

NO.2 导光控制面板 / Light guide control panel
机载、舰载、车载设备的控制与照明指示、一体化集成化设计
To realize the control and lighting instructions of aviation, carrier borne, vehicle mounted equipment. Integrated design. Various sizes can be customized

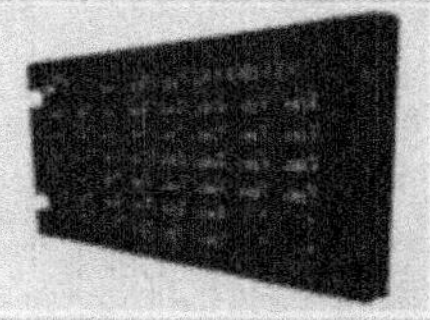

NO.3 军用触控产品 / Military touch products
超宽温储存与工作、抗声光干扰、电磁兼容性好
Extra wide temperature storage and work, Anti acoustic interference, visible light, High transmittance, good electromagnetic compatibility

NO.4 晶体、滤波器 / Crystal element
军用通信设备、雷达、电子对抗、试验及测量设备等
Communication equipment, radar, electronic countermeasure, test equipment

NO.5 被复线宽带综合数字远传通信设备 / By double broadband integrated digital remote communication equipment
士兵训练、部队作战
Soldier training Troops fighting

NO.6 军人心理智能测评及训练系统 / Soldier's Mental Status Intelligent Evaluating and Training
为飞行员或军人提供日常心理测评与训练
Provide routine mental test and training for pilots or soldiers

NO.7 安澜可信操作系统 / Alan Trusted Operating System
自主研发、自有知识产权、地址隐藏、加密芯片、证书认证、身份认证、安全认证机制
Independent R&D, and self-owned IP rights. Address hiding, encrypted chip, certificate authentication, identity authentication, security authentication mechanism

NO.8 智能手环 / AN-W1 AN-Watch
通过ZigBee私有网络与平板一对多组网通讯，保证了低功耗、短时延、高容量、高安全性
Tablet through ZigBee private network realize one-to-many networking communication, it ensures the low power consumption, short time delay, high capacity and high security

NO.9 北斗定位监控器 / Beidou Positioning Monitor
实时定位主设备的经纬度、海拔等信息，并校准时间，实现了远程监控
Positioning the latitude/longitude or altitude information of the major equipment at real time, and calibrate the time to realize remote monitoring

NO.10 飞行膝板 / Flight Knee Tablet
飞行员将膝板固定于腿部独立使用，便于携带与操作，专业高亮及加固设计
Pilot fixes the knee tablet on the leg for independent use to facilitate carrying and operation. Professional highlighting design and reinforcing design

NO.11 地图作业终端 / Map Operation Terminal
基于北斗卫星的地图作业系统
Map operation system based on Beidou satellite

NO.12 平板电脑 / Tablet
专业加固平板，防水防尘达到IP65等级，平板搭载了安澜可信操作系统
Professional reinforcing tablet, IP65 of anti-water and anti-dust level, Alan Trusted OS is installed on the tablet

NO.13 汉凡交互式电子手册 / Hanfan Interactive Electronic Manual
为复杂的军、民用设备提供动态拆卸、安装、操作、故障诊断与维护的交互式电子手册
An interactive electronic manual used to provide dynamic dismantle, installation, operation, trouble-shooting and maintenance to complicated military and folk equipment

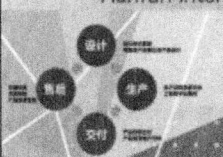

China Electronics Corp. (CEC). Booklet. Nanjing Huadong Electronics Group Corp., Ltd. (2/4). 2016 Zhuhai Airshow. (28/54).

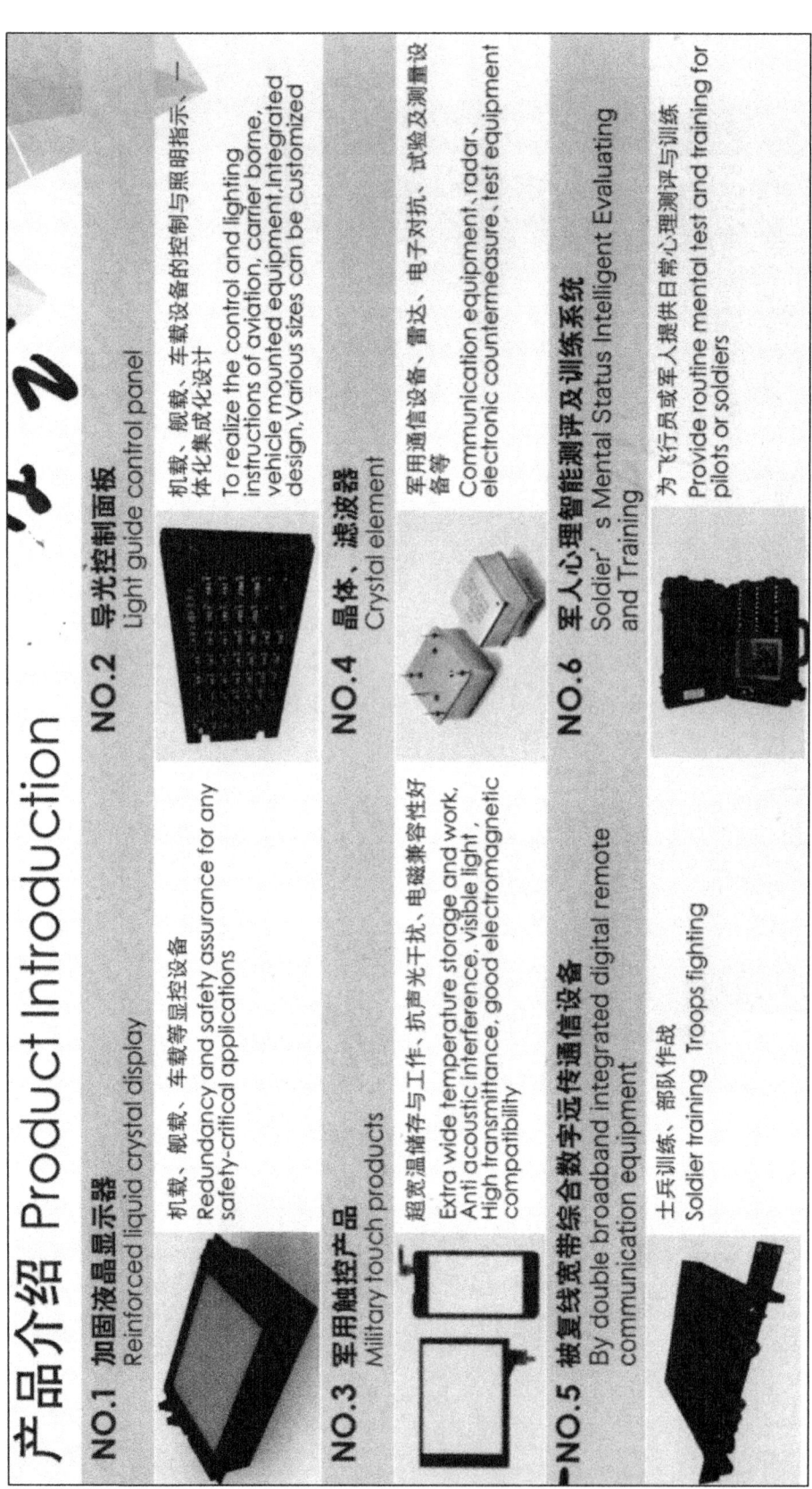

产品介绍 Product Introduction

NO.1 加固液晶显示器 Reinforced liquid crystal display

机载、舰载、车载等显控设备
Redundancy and safety assurance for any safety-critical applications

NO.2 导光控制面板 Light guide control panel

机载、舰载、车载设备的控制与照明指示一体化集成化设计
To realize the control and lighting instructions of aviation, carrier borne, vehicle mounted equipment, Integrated design, Various sizes can be customized

NO.3 军用触控产品 Military touch products

超宽温储存与工作、抗声光干扰、电磁兼容性好
Extra wide temperature storage and work, Anti acoustic inference, visible light, High transmittance, good electromagnetic compatibility

NO.4 晶体、滤波器 Crystal element

军用通信设备、雷达、电子对抗、试验及测量设备等
Communication equipment, radar, electronic countermeasure, test equipment

NO.5 被复线宽带综合数字远传通信设备 By double broadband integrated digital remote communication equipment

士兵训练、部队作战
Soldier training Troops fighting

NO.6 军人心理智能测评及训练系统 Soldier's Mental Status Intelligent Evaluating and Training

为飞行员或军人提供日常心理测评与训练
Provide routine mental test and training for pilots or soldiers

CLOSE-UP. China Electronics Corp. (CEC). Booklet. Nanjing Huadong Electronics Group Corp., Ltd. (3/4). 2016 Zhuhai Airshow. (29/54).

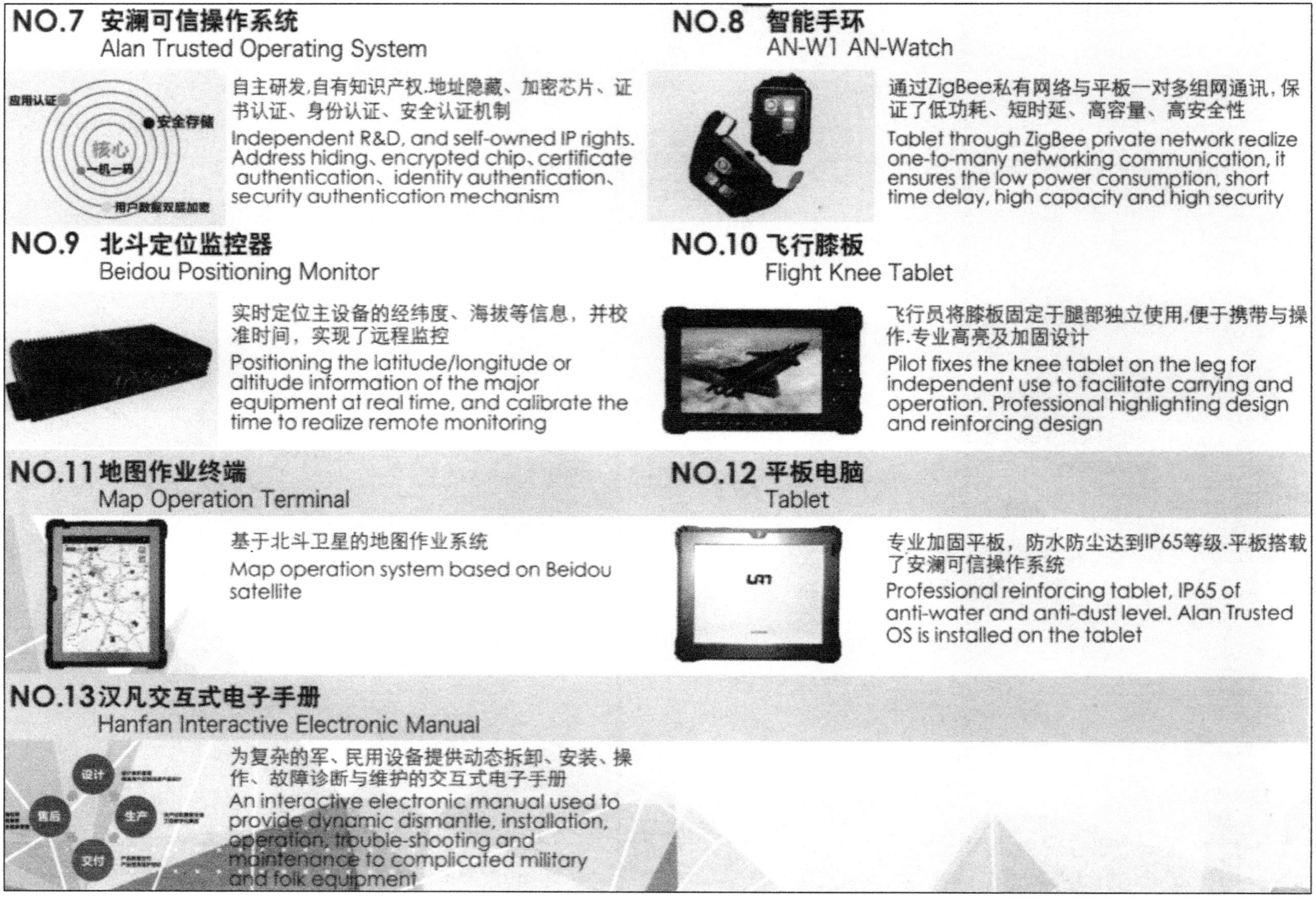

NO.7 安澜可信操作系统
Alan Trusted Operating System

自主研发,自有知识产权,地址隐藏,加密芯片、证书认证、身份认证、安全认证机制

Independent R&D, and self-owned IP rights. Address hiding, encrypted chip, certificate authentication, identity authentication, security authentication mechanism

NO.8 智能手环
AN-W1 AN-Watch

通过ZigBee私有网络与平板一对多组网通讯,保证了低功耗、短时延、高容量、高安全性

Tablet through ZigBee private network realize one-to-many networking communication, it ensures the low power consumption, short time delay, high capacity and high security

NO.9 北斗定位监控器
Beidou Positioning Monitor

实时定位主设备的经纬度、海拔等信息,并校准时间,实现了远程监控

Positioning the latitude/longitude or altitude information of the major equipment at real time, and calibrate the time to realize remote monitoring

NO.10 飞行膝板
Flight Knee Tablet

飞行员将膝板固定于腿部独立使用,便于携带与操作,专业高亮及加固设计

Pilot fixes the knee tablet on the leg for independent use to facilitate carrying and operation. Professional highlighting design and reinforcing design

NO.11 地图作业终端
Map Operation Terminal

基于北斗卫星的地图作业系统

Map operation system based on Beidou satellite

NO.12 平板电脑
Tablet

专业加固平板,防水防尘达到IP65等级,平板搭载了安澜可信操作系统

Professional reinforcing tablet, IP65 of anti-water and anti-dust level. Alan Trusted OS is installed on the tablet

NO.13 汉凡交互式电子手册
Hanfan Interactive Electronic Manual

为复杂的军、民用设备提供动态拆卸、安装、操作、故障诊断与维护的交互式电子手册

An interactive electronic manual used to provide dynamic dismantle, installation, operation, trouble-shooting and maintenance to complicated military and folk equipment

CLOSE-UP. China Electronics Corp. (CEC). Booklet. Nanjing Huadong Electronics Group Corp., Ltd. (4/4). 2016 Zhuhai Airshow. (30/54).

公司简介
Company Profile

长沙湘计海盾科技有限公司成立于2001年,拥有员工460余人,是专业从事军用计算机设备、军用显示设备、军用网络设备及光纤探测系统的研发、生产和售后服务企业,产品广泛装备于海陆空火箭军中。作为国内重要的军用航空显示设备研发和生产单位,研制的显控设备成功应用于"神舟飞船、天宫实验室"等国家重点型号和工程上。

Founded in 2001, Changsha Xiangji-Haidun Technology Co.,Ltd has more than 460 employees now. The company is a professional enterprise engaged in the research, development, production, after-sales service of military computer, military display, military network equipment and fiber optic detection system. The company's products are widely used in PLA Army, Navy, Air Force and Rocket Force. The company is the important research and production corporation of military aviation display equipment in China. The company's display and control system products is used in Shenzhou spacecraft, Tiangong spacelab and other national key projects.

Hiden 湘计海盾

地址: 湖南长沙经济技术开发区东三路5号　电话: 0731-84932900
传真: 0731-84932898　　　　　　　　　　　邮编: 410100
Add:No.5,DongSan Road,Economic development zone,Changsha,Hunan
Tel: 0731-84932898　　Fax:0731-84932898　　Postcode: 410100

China Electronics Corp. (CEC). Booklet. Changsha Xiangji-Haidun Technology Co., Ltd. (1/8). 2016 Zhuhai Airshow. (31/54).

LI15AFM型军用机载TFT液晶显示器
LI15AFM military airborne TFT LCD monitor

产品特点：

LI15AFM型军用机载TFT液晶显示器是满足座舱应用要求的大尺寸机载智能液晶显示器，开放式的软件架构和组合化结构，可灵活适应机载、航天显示要求。
- ◆支持多路外视频直接显示功能和融合、叠加显示；
- ◆具备运行数字地图软件显示能力（基本图元指令、渲染、填充、三维绘图能力）；
- ◆具有OpenGL图形开发功能，具有3D图形处理能力；
- ◆具有矢量汉字字库功能。

Product Features :

LI15AFM military airborne TFT LCD monitor is a large-size intelligent LCD, designed to meet the requirements of cabin airborne. Its software is of open architecture and modular structure, which makes it flexible to adapt to airborne and spaceship display requirements.
- ◆Support multi-way video direct, integration and overlay display;
- ◆Support digital map software display (Basic primitive Instructions, rendering, filling, 3D drawing capabilities, etc.);
- ◆Support OpenGL graphics development, being capable of processing 3D graphics;
- ◆Come with vector fonts of Chinese characters.

功能与性能：
Function and performance:

分辨率 Pixel format	1024×768 pixels 1024×768 pixels
亮度 Brightness	16.7M 16.7M
色域 Color gamut	≥1000cd/m2 ≥1000cd/m2
对比度 Contrast	500:1 500:1
视角 Active area	150° (H), 120° (V) 150° (H), 120° (V)
电源 Power	DC28V DC28V
功耗 Power consumption	≤50W,低温≤120W ≤50W, Low temperature ≤120W
接口 Interface	VGA、DVI、LVDS、PAL1553B、ARINC429、100M以太网、RS-422、IEEE 1394 VGA, DVI, LVDS, PAL1553B, ARINC429, 100M Ethernet, RS-422, IEEE 1394

环境适应性：
Environmental adaptability:

工作温度 Operating temperature	-45℃~60℃ -45℃~60℃
贮存温度 Storage temperature	-55℃~70℃ -55℃~70℃
振动 Vibration	满足GJB150.16-1986机载要求 Meet the airborne requirements of GJB150.16-1986
冲击 Impact	满足GJB150.18-1986机载要求 Meet the airborne requirements of GJB150.16-1986
电磁兼容 EMC	满足GJB151A-1997相关要求 Meet the related requirements of GJB151A-1997

China Electronics Corp. (CEC). Booklet. Changsha Xiangji-Haidun Technology Co., Ltd. (2/8). LI15AFM military Airborne TFT LCD2016 Monitor. Zhuhai Airshow. (32/54). Apologize for the black mark across page. See CLOSE-UP on next pages.

LI15AFM型军用机载TFT液晶显示器
LI15AFM military airborne TFT LCD monitor

产品特点：

LI15AFM型军用机载TFT液晶显示器是满足座舱应用要求的大尺寸机载智能液晶显示器，开放式的软件架构和组合化结构，可灵活适应机载、航天显示要求。
- 支持多路外视频直接显示功能和融合、叠加显示；
- 具备远行数字地图软件显示能力（基本图元指令、渲染、填充、三维绘图能力）；
- 具有OpenGL图形开发功能，具有3D图形处理能力；
- 具有矢量汉字字库功能。

Product Features :

LI15AFM military airborne TFT LCD monitor is a large-size intelligent LCD, designed to meet the requirements of cabin airborne. Its software areas of open architecture and modular structure, which makes it flexible to adapt to airborne and spaceship display requirements.
- Support multi-way video direct, integration and overlay display;
- Support digital map software display (Basic Primitive Instructions, rendering, filling, 3D drawing capabilities, etc.);
- Support OpenGL graphics development, being capable of processing 3D graphics;
- Come with vector fonts of Chinese characters.

CLOSE-UP: China Electronics Corp. (CEC). Booklet. Changsha Xiangji-Haidun Technology Co., Ltd. (3/8). LI15AFM military Airborne TFT LCD2016 Monitor. 2016 Zhuhai Airshow. (33/54).

功能与性能：
Function and performance:

分辨率 Pixel format	1024×768 pixels 1024×768 pixels
亮度 Brightness	16.7M 16.7M
色域 Color gamut	≥1000cd/m2 ≥1000cd/m2
对比度 Contrast	500:1 500:1
视角 Active area	150° (H) , 120° (V) 150° (H) , 120° (V)
电源 Power	DC28V DC28V
功耗 Power consumption	≤50W,低温≤120W ≤50W, Low temperature ≤120W
接口 Interface	VGA、DVI、LVDS、PAL1553B、ARINC429、100M以太网、RS-422、IEEE 1394 VGA、DVI、LVDS、PAL1553B、ARINC429、100M Ethernet、RS-422、IEEE 1394

环境适应性：
Environmental adaptability:

工作温度 Operating temperature	-45℃～60℃ -45℃～60℃
贮存温度 Storage temperature	-55℃～70℃ -55℃～70℃
振动 Vibration	满足GJB150.16-1986机载要求 Meet the airborne requirements of GJB150.16-1986
冲击 Impact	满足GJB150.18-1986机载要求 Meet the airborne requirements of GJB150.16-1986
电磁兼容 EMC	满足GJB151A-1997相关要求 Meet the related requirements of GJB151A-1997

CLOSE-UP: China Electronics Corp. (CEC). Booklet. Changsha Xiangji-Haidun Technology Co., Ltd. (4/8). LI15AFM military Airborne TFT LCD2016 Monitor. 2016 Zhuhai Airshow. (34/54).

机载图形板卡设备
Airborne Graphics Card Equipment

产品特点：

结构特点：标准6U CPCI板卡，无风扇设计
性能：最大4路独立输出，最大分辨率2560*1600·60HZ
功耗：不大于30W，典型工耗25W
CPU：Freescale PowerPC P1022/T1024，双核主频1G
GPU：AMD E4690/E6465
系统内存2GB/667MHZ，显存512MB/700MHZ、2048MB/3200MHZ
系统：vxWorks6.9操作系统；
接口：显示输出2路VGA、2路DVI、2路LVDS、1路HDMI、1路DP；可同时独立输出2~4路视频；具备2路RS232、1路100M以太网通讯接口；OpenGL1.3/OpenGL2.0
温度：工作温度-40℃~+85℃；

产品应用：

该设备主要为各型飞机提供高性能的2D、3D图像显示，可支持多路2K分辨率图像输出，采用无风扇设计，以powerPC处理器及嵌入式GPU为硬件平台，运行在vxWorks操作系统上，具有可靠性高、功耗低、耐高温等特点，适用恶劣环境下，高性能2D、3D图象显示。

Product Features :

Structure: Standard 6U CPCI Card with fanless design;
Capability: Maximum 4 independent output, Maximum resolution 2560*1600·60HZ
Power dissipation: <30W, typical 25W
CPU: Freescale PowerPC P1022/T1024, Dual core, Basic Frequency 1GHZ
GPU: AMD E4690/E6465
System Memory 2GBytes/667MHZ, VRAM 512MBytes/700MHZ、2048MB/3200MHZ
Operation System: vxWorks6.9
Interfaces: 2 VGA、2 DVI、2 LVDS、1 HDMI and 1 DP video output; simultaneously and independently output 2~4 video signals; 2 RS232、1 100MEthernet communication interface; OpenGL1.3/OpenGL2.0
Temperature: Operating temperature -40℃~+85℃;

Application :

This equipment is mainly used for the various types of aircrafts with high-performance 2D and 3D image display, which support multiple 2K resolution image output. It adopts the fanless design, and is based on powerPC processor and embedded GPU hardware platform, working on vxWorks operating system. It is of high reliability, low power consumption, high temperature resistance, and thus, suitable for high-performance 2D and 3D display in adverse environment.

China Electronics Corp. (CEC). Booklet. Changsha Xiangji-Haidun Technology Co., Ltd. (5/8). Airborne Graphics Card Equipment. 2016 Zhuhai Airshow. (35/54). Apologize for the black mark across page. See CLOSE-UP on next pages.

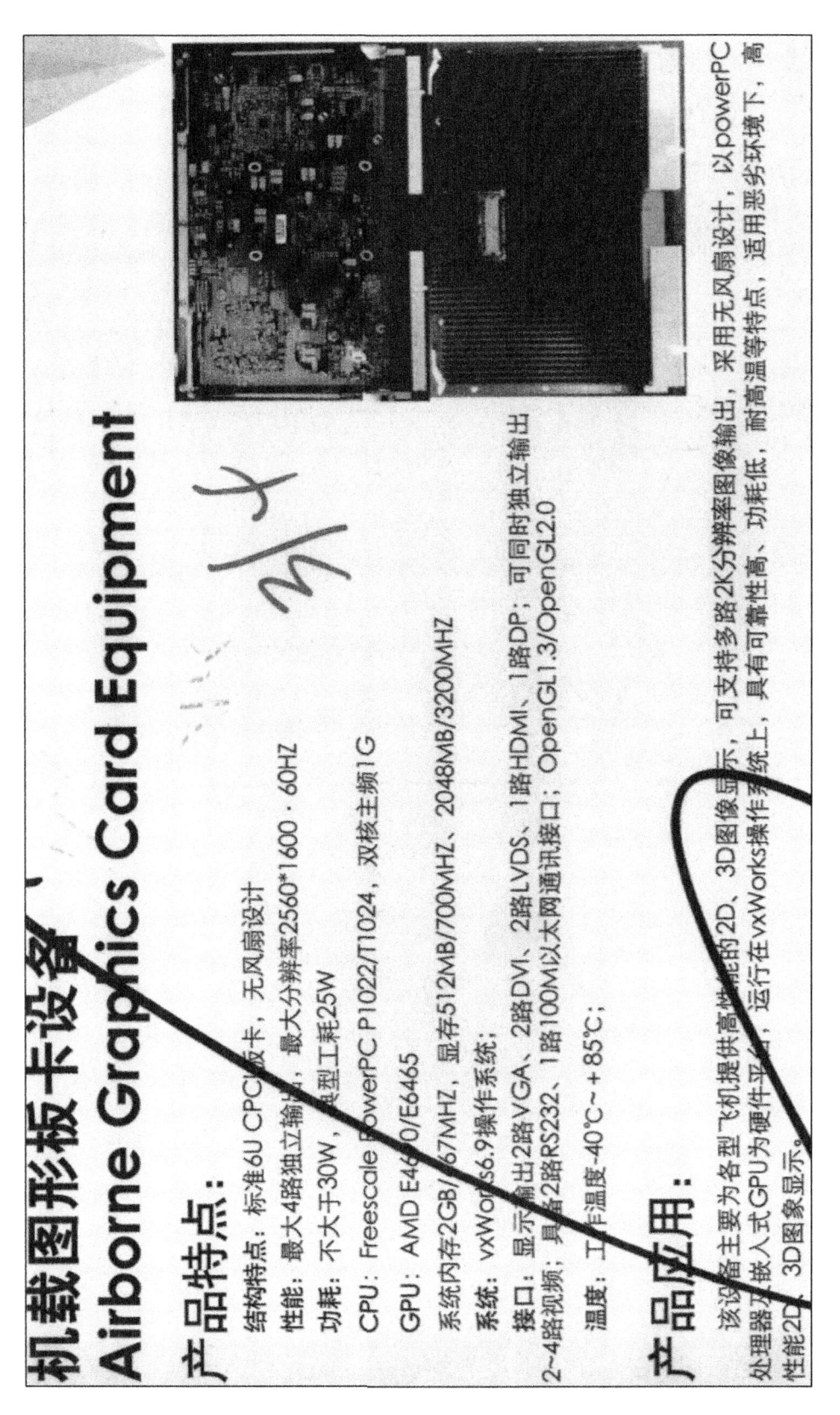

机载图形板卡设备
Airborne Graphics Card Equipment

产品特点：

- 结构特点：标准6U CPCI板卡，无风扇设计
- 性能：最大4路独立输出，最大分辨率2560*1600，60HZ
- 功耗：不大于30W，典型工耗25W
- CPU：Freescale PowerPC P1022/T1024，双核主频1G
- GPU：AMD E4690/E6465
- 系统内存2GB/667MHZ，显存512MB/700MHZ，2048MB/3200MHZ
- 系统：vxWorks6.9操作系统；
- 接口：显示输出2路VGA、2路DVI、2路LVDS、1路HDMI、1路DP；可同时独立输出 2~4路视频；具备2路RS232、1路100M以太网通讯接口；OpenGL1.3/OpenGL2.0
- 温度：工作温度-40℃~+85℃；

产品应用：

该设备主要为各型飞机提供高性能的2D、3D图像显示，可支持多路2K分辨率图像输出，以powerPC处理器及嵌入式GPU为硬件平台，运行在vxWorks操作系统上，具有可靠性高、功耗低、耐高温等特点，适用恶劣环境下，高性能2D、3D图象显示。

Close-Up: China Electronics Corp. (CEC). Booklet. Changsha Xiangji-Haidun Technology Co., Ltd. (6/8). Airborne Graphics Card Equipment. 2016 Zhuhai Airshow. (36/54).

Product Features:

Structure: Standard 6U CPCI Card with fanless design;
Capability: Maximum 4 independent output, Maximum resolution 2560*1600 · 60HZ
Power dissipation: <30W, typical 25W
CPU: Freescale PowerPC P1022/T1024, Dual core, Basic Frequency 1GHZ
GPU: AMD E4690/E6465
System Memory 2GBytes/667MHZ, VRAM 512MBytes/700MHZ、2048MB/3200MHZ
Operation System: vxWorks6.9
Interfaces: 2 VGA、2 DVI、2 LVDS、1 HDMI and 1 DP video output; simultaneously and independently output 2~4 video signals; 2 RS232, 1 100MEthernet communication interface; OpenGL1.3/OpenGL2.0
Temperature: Operating temperature -40°C～+85°C;

Application:

This equipment is mainly used for the various types of aircrafts with high-performance 2D and 3D image display, which support multiple 2K resolution image output. It adopts the fanless design, and is based on powerPC processor and embedded GPU hardware platform, working on vxWorks operating system. It is of high reliability, low power consumption, high temperature resistance, and thus, suitable for high-performance 2D and 3D display in adverse environment.

Close-Up: China Electronics Corp. (CEC). Booklet. Changsha Xiangji-Haidun Technology Co., Ltd. (7/8). Airborne Graphics Card Equipment. 2016 Zhuhai Airshow. (37/54).

携行以太网交换机
Portable Ethernet Switch

产品特点：

携行以太网交换机，具有便携、智能、多制式传输、远距离通信、电池长时间续航运行等特点。

接口：15个10/100Mbps自适应

　　　1个10/100/1000Mbps自适应

　　　1对远传接口

功耗：不大于10W

远传功能：最远传输距离可达8km，通信制式、传输速率可调

续航能力：13小时

重量：4.5Kg

尺寸：270mm（长）×255mm（宽）×100mm（高）

工作温度：-40℃~+60℃

产品应用：

广泛应用于多军兵种的区域情报处理系统、指挥系统、数据链系统，有快速组网要求的环境中。

Product Features :

Portable Ethernet switch is portable and intelligent, which provides multi-standard transmission, long-distance communication, and long battery endurance.

Interface: 15 ports 10/100 Mbps AUTO Negotiation

　　　　　1 ports 10/100/1000 Mbps AUTO Negotiation

　　　　　A pair of long distance transmission interface

Power dissipation: ≤10W

Long distance transmission: up to 8km with adjustable communication mode and transmission rate.

Battery endurance: 13 hours

Weight: 4.5Kg

Size: 270mm (L) ×255mm (W) ×100mm (H)

Operating temperature: -40℃~+60℃

Application :

It is widely used in regional intelligence processing systems, command systems, data link systems, and environment that requires fast network deployment.

Hiden 湘计海盾

地址：湖南长沙经济技术开发区东三路5号　电话：0731-84932900
传真：0731-84932898　邮编：410100
Add:No.5,DongSan Road,Economic development zone,Changsha,Hunan
Tel: 0731-84932898　Fax: 0731-84932898　Postcode: 410100

China Electronics Corp. (CEC). Booklet. Changsha Xiangji-Haidun Technology Co., Ltd. (8/8). 2016 Zhuhai Airshow. (38/54).

中软信息系统工程有限公司
CS&S Informatica System Engineering Co.,Ltd

中软信息系统工程有限公司是中国电子下属核心军工科研企业,纳入国家军工计划管理渠道。公司是中国电子信息安全业务板块的核心骨干企业,长期致力于网络信息安全和国家、国防重要行业领域信息化建设。

CS&S is the CEC subordinate core military scientific research enterprise. It is included in the national military program management channels. The company is the backbone of the information security business segment of the CEC backbone enterprises. It is committed to long-term network information security and national, national defense important industry information construction.

主营业务
Main Business

自主可控体系架构及平台应用技术研究	大型复杂信息系统集成	软件开发	咨询服务
self-controllable architecture and platform application technology research	large-scale complex information systems integration	software development	consulting service

下设单位
Under the Subsidiary

公司下设迈普通信技术股份有限公司(控股)和自主可控软硬件联合攻关基地(国家级)。
CS&S consists of Maipu Communication Technology Co., Ltd.(Holdings) and independent control software and hardware joint research base (national).

地址:北京市昌平区北七家镇未来科技城南区中国电子网络安全与信息化产业基地D座5层
Add: Floor 5, Building D, CEC Industry Base, the future of science and technology south area, Beiqila, Changping District, Beijing

电话 Tel: +86010-56956666
传真 Fax: +86010-56956688
邮箱 Mail: luzhen@css.com.cn

China Electronics Corp. (CEC). Booklet. CS&S Informatica System Engineering, Co., Ltd./Maipu Communication Technology Co., Ltd. (1/1). 2016 Zhuhai Airshow. (39/54).

迈普通信技术股份有限公司
Maipu Communication Co., Ltd. (Maipu)

2016 Zhuhai

公司简介 Company Profile

迈普是中国电子（CEC）旗下企业，创立于1993年，是国内主流、拥有自主知识产权的基础网络设备及行业应用服务提供商。成立以来，迈普坚持走自主创新、产业报国之路，专注服务于关系国计民生的国防、金融、运营商、财政、税务、公安、电力、能源等行业，以专精深厚实的态度赢得了行业用户和社会各界的认可。

Maipu, owned enterprise of CEC, founded in 1993, is the mainstream basic network equipment and industry application service provider with independent intellectual property rights in China. Since its inception, Maipu adheres to the road of serving the country by developing the industry based on the independent innovation, and is dedicated to the industries related with the nation's economy and the people's livelihood, such as defense, finance, business, finance, taxation, public security, electricity, and energy. With the professional and profound attitude, Maipu wins the appreciation of the industry users and social circles.

MAIPU 迈普

地址：成都市高新区九兴大道16号迈普大厦，邮编：610041，总机：(+86) 028-85148048，传真：(+86) 028-85148948
Add: No.16, Jiuxing Avenue, High-tech Park, Chengdu, Sichuan, P.R.China, Zip code:610041
Tel:(+86) 028-85148048, Fax:(+86) 028-85148948, Website : http://www.maipu.cn

China Electronics Corp. (CEC). Booklet. Maipu Communication Technology Co., Ltd. (1/2). NSS6600; NSR2900. 2016 Zhuhai Airshow. (40/54).

迈普自主可控交换机 NSS6600
Maipu self-controllable switch NSS6600

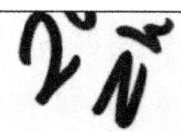

MyPower NSS6600自主可控万兆核心交换机是迈普公司面向政府、军工等安全性要求较高的行业推出的基于国产CPU、国产交换芯片的新一代多业务高性能的以太网交换产品。主要针对园区网/企业网的核心、数据中心汇聚等场景应用，提供从芯片到硬件到软件的全方位安全可控、稳定、可靠的高性能L2/L3层交换服务。

MyPower NSS6600 self-controllable Gigabit core switch is the new-generation multi-service high-performance Ethernet switch product developed by Maipu based on the domestic CPU and exchange chip for the industries with high requirement for the security, such as government and military. It provide a full range of safe, controllable, stable and reliable high-performance L2/L3 switching services from the chip to the hardware and software mainly for the campus network / enterprise network core, data center aggregation and other scenarios.

迈普自主可控路由器 NSR2900
Maipu self-controllable domestic router NSR2900

NSR2900自主可控路由器以国产多核处理器为核心处理器件，配套迈普公司自主设计研发并稳定运行十多年的网络操作系统及应用软件。全面支持IPv4、IPv6、MPLS等网络协议及访问控制、攻击检测等安全特性，从软硬件系统设计层面全面保证设备的可控性、安全性，为中国建设自主的安全网络从"芯"定制安全可靠的技术解决方案。

NSR2900 self-controllable domestic router takes the domestic multi-core processor as the core processor and cooperates with the network operation system and application software developed by Maipu and running for more than ten years independently. It fully supports IPv4, IPv6, MPLS and other network protocols, and access control, attack control and other security features, fully ensuring the controllability and security of the device from the software and hardware system design, and customizing the safe and reliable technical solutions from the core for China to construct the safe network by self.

China Electronics Corp. (CEC). Booklet. Maipu Communication Technology Co., Ltd. (2/2). 2016 Zhuhai Airshow. (41/54).

天津麒麟信息技术有限公司
Tianjin Kylin Information Technology Co., Ltd.

天津麒麟信息技术有限公司是在中国电子集团、国防科技大学和天津市政府支持下成立的国有控股信息技术高科技企业。

公司旗下的"银河麒麟"操作系统是国内安全等级最高的操作系统，源自国家"十五"规划科技重大专项的研究成果。经过十余年的发展，形成了服务器、桌面、实时、云、存储五大系列操作系统产品，以及高性能计算、云计算和云桌面等产品。"银河麒麟"已成功应用于"天河"超级计算机，并在国防、政务、电力、金融、能源、教育等行业得到广泛应用，成为我国安全可控信息系统的坚强基石。

公司秉承"军民融合、产业协作、国际合作"的理念，拥有强大稳定的核心技术团队，在国际开源社区具备重要影响力，将逐步发展成为世界级系统软件领先企业。

Tianjin Kylin Information Technology Co., Ltd. is a China famous high technology company which focus on providing leading Enterprise Linux Operating system, high performance Cloud computing and ARM CPU eco-system development. Tianjin Kylin headquarter locates in Tianjin, sales and marketing center in Beijing and R&D centers in Changsha, Tianjin as well.

Kylin Enterprise Linux OS is providing the highest level of security protection service of system software products, including Kylin Enterprise Linux Server Series/Kylin Enterprise Linux Desktop Series, Kylin-cloud computing, cloud desktops and solutions.

Tianjin Kylin has a strategy cooperation with Phytium company whom is the first ARM CPU provider for enterprise application, and Kylin Enterprise Linux OS is the only Linux OS provider supporting FT CPU.

Kylin OS series products have been widely applied in most industries including E-government, Finance, Energy resources, Public security, State Grid electricity ,etc., "Kylin OS+ FT CPU(ARM64)" have been successfully accepted by China Enterprise customers because of higher performance and better compatibility in ARM Eco-Chain application.

Our Mission is to be the best Arm based Enterprise Linux OS provider in China market, taking great effort to grow the ARM ecosystem in China.

KYLIN 银河麒麟
地址：天津市滨海新区海洋高新区信安创业广场3号楼　电话：86-022-58955650
传真：86-022-58955651　网址：www.kylinos.cn
邮箱：market@kylinos.cn　微信公众号：天津麒麟

China Electronics Corp. (CEC). Booklet. Tianjin Kylin Information Technology Co. Ltd. (1/4). 2016 Zhuhai Airshow. (42/54).

银河麒麟服务器操作系统
Kylin Server Operating System

银河麒麟服务器操作系统是在国家"863计划"和国家发改委产业化专项支持下，研制而成的强安全、高可靠、高可用国产操作系统，已在国防、军工、政务、电力、航天、金融、电信、教育、大中型企业等行业或领域得到广泛应用。

Kylin server operating system is in the national "863 Program" and the National Development and Reform industrialization special support from the development of strong security, high reliability, high availability domestic operating system, has been in defense, military, government, power, aerospace, finance, telecommunications, education, and medium-sized enterprises, and other industries or fields are widely used.

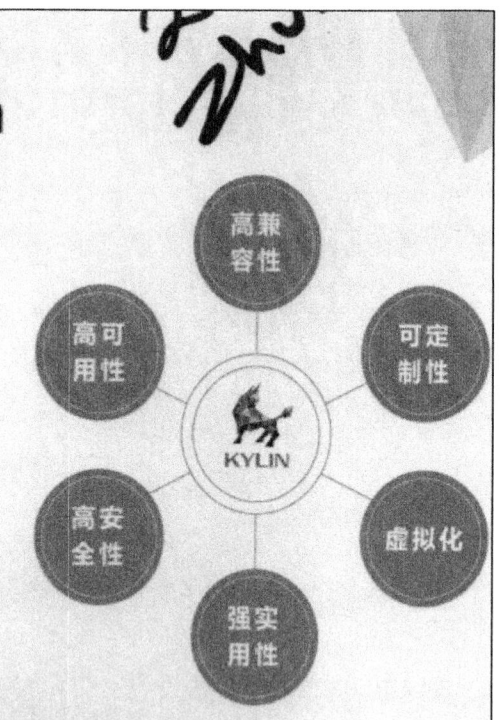

银河麒麟桌面操作系统
Kylin Desktop Operating System

银河麒麟桌面操作系统是软硬件兼容性最好的国产桌面操作系统，拥有绚丽的人机交互界面，友好易用，用户十分钟便可轻松掌握。银河麒麟桌面操作系统主要面向电子办公、家庭生活、个人娱乐。中文与办公具有良好的中文安装界面及操作界面，默认集成搜狗输入法，支持WPS、永中Office等办公软件。

Kylin desktop operating system is the best-made hardware and software compatibility desktop operating system, has a brilliant man-machine interface, friendly and easy to use, the user can easily grasp ten minutes. Kylin desktop operating system mainly for electronic office, family life, personal entertainment. Chinese office and has a good Chinese installation interface and user interface, the default integrated Sogou input method support WPS, Evermore Office and other office software.

China Electronics Corp. (CEC). Booklet. Tianjin Kylin Information Technology Co. Ltd. (2/4). Kylin Server Operating System; Kylin Desktop Operating System. 2016 Zhuhai Airshow. (43/54).

银河麒麟云桌面管理系统
Kylin Cloud Desktop Management System

银河麒麟云桌面管理系统是基于银河麒麟云的成熟完善的企业级桌面虚拟化平台。银河麒麟云桌面将操作系统、应用、数据和配置文件从底层硬件中分离出来，集中在云中心，实现对桌面/应用/数据的统一管理、统一存储和统一计算，向用户终端交付虚拟桌面/应用等服务，确保本地不留密、网络不传密，简化桌面系统的维护和管理。

Kylin Cloud desktop management system is based on the Kylin cloud sophisticated enterprise-class desktop virtualization platform. Kylin cloud desktop operating system, applications, data and configuration files separate from the underlying hardware out, focused on the cloud center, to achieve desktop / application / data unified management, unified storage and unified computing, to the user terminal delivering virtual desktops / applications and services, to ensure that local does not stay secret, secret network does not pass, simplify desktop management and system maintenance.

银河麒麟高性能计算集群系统
Kylin HPC Cluster System

银河麒麟高性能计算系统是在"银河"/"天河"系列超级计算机研究成果的基础上开发的适合中小规模应用需求的高性能计算系统。系统采用了自主研发的银河麒麟操作系统和银河麒麟高性能计算（HPC）套件，支持InfiniBand、万兆高速计算网络，支持国内外主流的高性能服务器、刀片服务器、海量存储、高速网络交换等硬件设备，提供资源管理、HPC集群管理、并行编译、并行计算等功能，主要面向密码算法、卫星遥感、气象预报、生物医药、资源勘测、图像处理等领域，满足科学计算和海量数据处理。

Kylin system is based on high-performance computing "Galaxy" / "Milky Way" series supercomputer research or the development of high-performance computing systems for small and medium-scale application needs. The system uses a self-developed Kylin Kylin operating system and high-performance computing (HPC) suite that supports InfiniBand, Gigabit high-speed computing network to support domestic and international mainstream high-performance servers, blade servers, mass storage, and other high-speed network switching hardware equipment, resources management, HPC cluster management, parallel compilation, parallel computing, and other functions, mainly for cryptographic algorithms, satellite remote sensing, weather forecasting, bio-medicine, resource survey, image processing and other fields to meet the massive data processing and scientific computing.

China Electronics Corp. (CEC). Booklet. Tianjin Kylin Information Technology Co. Ltd. (3/4). Kylin Cloud Desktop Management System; Kylin HPC Cluster System. 2016 Zhuhai Airshow. (44/54).

银河麒麟云平台管理系统
Kylincloud Platform Management System

银河麒麟云是以模块化、可插拔为设计理念,可按需使用、易于管理、灵活扩展、安全可靠的新一代云平台管理系统。

银河麒麟云采用高效的资源池化机制,支持多种主流虚拟化方式,通过网络将IT基础设施资源按需提供给用户使用(IaaS服务),同时支持PaaS服务和SaaS服务。

银河麒麟云采用云-端一体化设计,支持用户对资源进行远程桌面访问,通过易用高效的远程访问协议,获得接近本地使用的体验。

银河麒麟云致力于为用户提供安全、弹性、高可用、高性能的公有云/私有云/专业云解决方案。

KylinCloud in a modular, pluggable for the design, on-demand use, easy to manage, flexible, scalable, secure and reliable next-generation cloud platform management system.

Kylin cloud using an efficient resource pooling mechanism to support a variety of mainstream virtualization way through the IT network infrastructure resources available to users on demand (IaaS services), support services and SaaS PaaS service.

Kylin cloud cloud - end integrated design that allows users to perform remote desktop access to resources, through easy and efficient remote access protocol, get close to the local use experience.

Kylin cloud to provide users with secure, flexible, high availability, high-performance public cloud / private cloud / Professional cloud solutions.

银河麒麟云操作系统
KylinCloud Operating System

银河麒麟云操作系统是天津麒麟面向军队、政府、金融、交通、电力、电信、医疗等多个行业客户推出的全新国产自主可控操作系统,提供全虚拟、半虚拟和容器虚拟化,全面支持国产飞腾CPU和银河麒麟云(KylinCloud),作为云计算支撑平台支持多种类型业务应用,适用于数据中心及云计算中心。

KylinCloud operating system is cloud-oriented industries in Tianjin Kylin military, government, finance, transportation, electricity, telecommunications, medical and other domestic customers launched a new self-controlled operating system, providing full virtualization, para-virtualized and virtualized containers full support Phytium CPU and Kylin cloud (KylinCloud), as a cloud computing platform to support multiple types of services to support applications for data center and cloud computing center.

China Electronics Corp. (CEC). Booklet. Tianjin Kylin Information Technology Co. Ltd. (4/4). Kylincloud Platform Management System; KylinCloud Operating System. 2016 Zhuhai Airshow. (45/54).

中标软件有限公司
China Standard Software Co., Ltd.

公司简介 Company Profile

中标软件有限公司是中国Linux操作系统和办公软件的提供商和服务商。作为国家规划布局内重点软件企业，中标软件通过了CMMI5认证，获得了国防和民用企业与产品资质。中标麒麟操作系统产品荣获"国家重点新产品"，2011-2015连续五年位列中国Linux操作系统市场占有率第一。中标软件成为在品牌、市场、业务实力等方面国内领先的操作系统旗舰企业。

China Standard Software Co., Ltd. (CS2C) is a domestic Linux Operating System(OS) and Office Software supplier. As a key company within the state software industrial layout, CS2C has got the CMMI5 certification and many other qualifications from national defense and civil government. NeoKylin OS won the honor of 'national key new products'. From 2011 to 2015, Neokylin continuously ranked the No.1 in China Linux OS market. CS2C has become the leading OS enterprise in China just because its high quality brands, broad market and powerful technical strength.

地址：上海市徐汇区番禺路1028号10楼　电话:021-51098866　传真:021-51062866
Add: F10 Shuyu Building, No.1028 Panyu Road, Shanghai (200030) China
Tel: 021-51098866　Fax: +86 21-51062866　Website: www.cs2c.com.cn

China Electronics Corp. (CEC). Booklet. China Standard Software Co. Ltd. (1/5). Neokylin Operating System. 2016 Zhuhai Airshow. (46/54).

中标麒麟操作系统
Neokylin Operating System

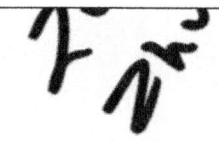

目前，中标麒麟操作系统已经在政府、国防、金融、教育、财税、公安、审计、交通、医疗、制造等行业得到深入应用，2011-2015年，连续五年位列中国Linux操作系统市场占有率第一位，中标软件也连续获得国防、民用领域核高基重大专项支持。

中标麒麟操作系统作为国产操作系统旗舰，以可信安全操作系统技术为重点，打造完善的产业生态，全面支持X86、ARM平台，以及龙芯、申威、兆芯、众志、华为等国产芯片，并实现了代码同源，用户体验一致。

As domestic flagship, the Neokylin Operating System focuses on the technology of trusted OS. In order to create a completed industrial ecology, Neokylin can not only support X86, ARM platforms, but also support Loongson, Sunway, Zhaoxin, MPRC, HUAWEI platforms well. The series OS products of Neokylin came from the same source code, their user fexperiences are consistent.

Now, the Neokylin Operating System has been widely used in many fields, such as government, national defense, finance, education, taxation, public security, auditing, transportation, medical, manufacturing etc. From 2011 to 2015, Neokylin continuously ranked the No.1 in China Linux OS market. Our company has also been supported by national defense and government through major national projects.

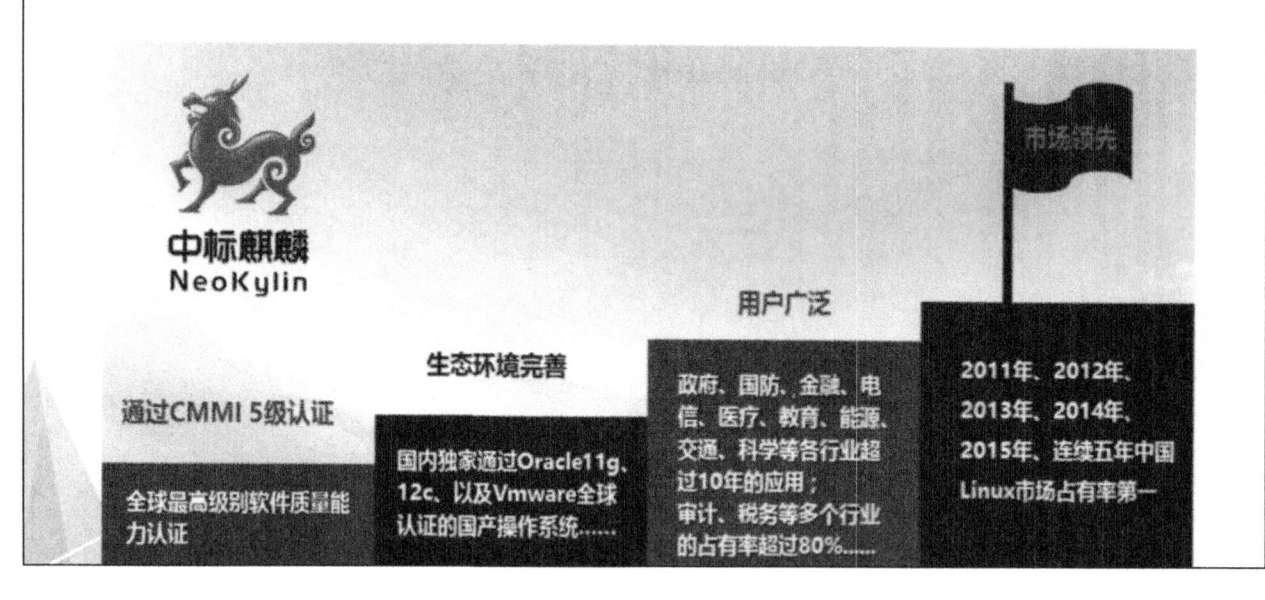

China Electronics Corp. (CEC). Booklet. China Standard Software Co. Ltd. (2/5). NeoKylin HA Cluster. 2016 Zhuhai Airshow. (47/54).

中标麒麟高可用集群软件

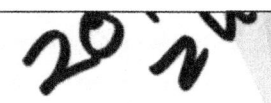

中标麒麟高可用集群软件致力于为用户打造业务连续性保障、数据持续保护、灾难恢复的高可用环境。中标麒麟高可用集群软件利用健康检测、磁盘心跳、秒级切换等功能，解决了软硬件及人为原因造成的单点/集群故障而引起的业务中断，有效确保单点系统或集群上关键任务应用程序和数据的稳定性和可靠性，为政府、金融、电力、医疗、运输、制造业等行业的用户提供高效、至微的可靠服务。

NeoKylin HA Cluster is committed to building for users HA environment featuring sustaining business guarantee, data protection and disaster recovery. Utilizing many functions like health monitoring, disk heartbeat and second level switch, NeoKylin HA Cluster settled the problem of single-point and cluster malfunction caused by hardware and softwares as well as human error. It ensures major applications and data in single-point and cluster stable and reliable, providing better service in fields of government, finance, electricity, medicine, transportation and manufacturing.

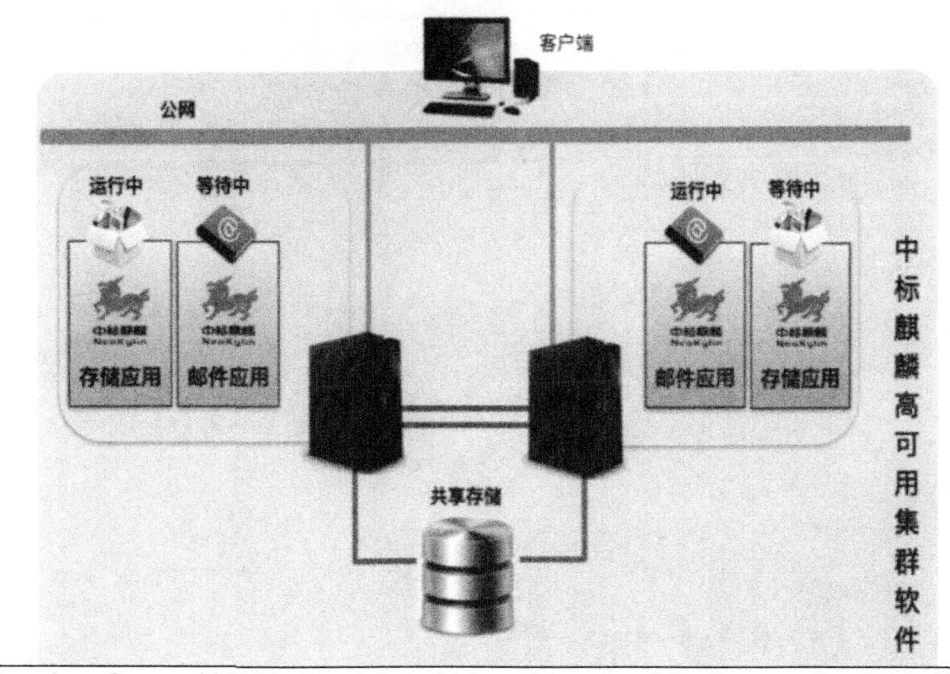

China Electronics Corp. (CEC). Booklet. China Standard Software Co. Ltd. (3/5). NeoKylin HA Cluster. 2016 Zhuhai Airshow. (48/54).

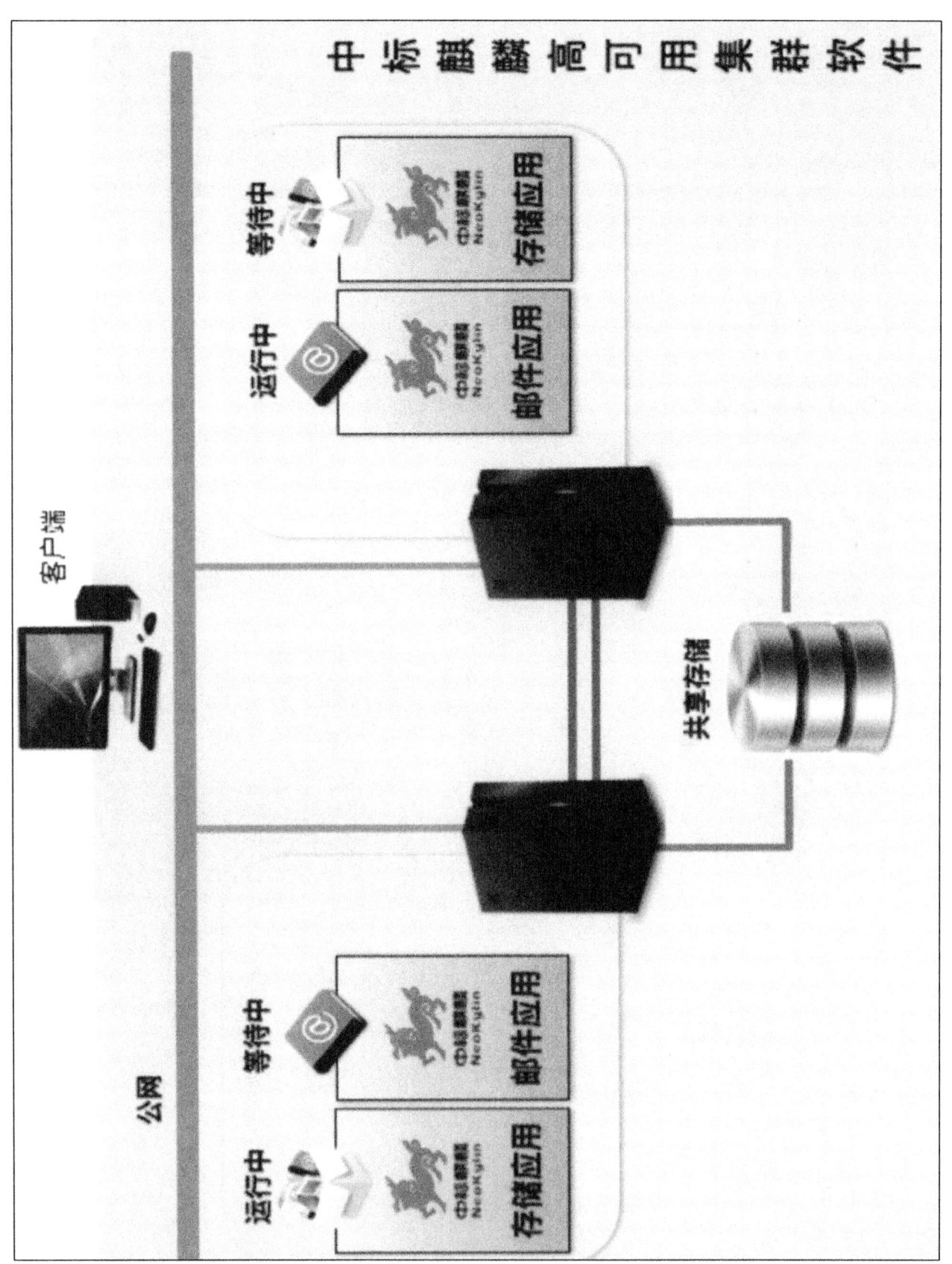

CLOSE-UP: China Electronics Corp. (CEC). Booklet. China Standard Software Co. Ltd. (4/5). NeoKylin HA Cluster. 2016 Zhuhai Airshow. (49/54).

中标普华办公软件
Neoshine Office

中标普华办公软件是面向我国政府和企业应用推出的一款稳定、可靠、性能最优及可定制化的办公软件产品，能够跨平台稳健运行，双向兼容MS Office等主流办公软件。该产品以满足中国办公市场需求为出发点，深入挖掘国外办公软件不能满足的本地化应用需求，在涵盖办公软件常用功能的基础上，进一步提供符合中国办公需求的特色功能，形成了具有中国特色的办公软件。

Neoshine Office is a stable, reliable, performance optimal and customizable office software product launch for the Chinese government and enterprise application, which is able to cross-platform and two-way compatible mainstream office software like MS Office. This product to meet the needs of Chinese office market as the starting point, in-depth localization of the application requirements that foreign office software cannot meet, based on covering the commonoffice software function, further provide Chinese demand for office features, forming the office software with China characteristic.

China Electronics Corp. (CEC). Booklet. China Standard Software Co. Ltd. (5/5). NeoShine Office. 2016 Zhuhai Airshow. (50/54).

中国电子信息产业集团有限公司第六研究所
The 6th Research Institute of China Electronics Corporation

公司简介 Company Profile

中国电子信息产业集团有限公司第六研究所（又名华北计算机系统工程研究所，简称电子六所）成立于1965年，直属中国电子信息产业集团有限公司（CEC），我国最早从事电子技术应用系统研究、开发的重点科研院所之一，设有工业控制系统信息安全国家工程实验室。电子六所围绕工控系统及安全、高新电子、现代信息服务三大主业，致力于打造工控系统及安全领域国内领先企业、国内一流技术水平的军民融合示范企业、国防领域具有战略地位的关键科研机构。

The 6th Research Institute was established in 1965, and is affiliated with the China Electronics Corporation. One of the earliest institutes to engage in developing and researching electronic technology application systems, the 6th Research Institute is the national engineering laboratory for industrial control systems. Our institute operates mainly in three business areas: industrial control systems, high tech electronics and modern information services. The institute is focused on establishing a first-class army-civilian integrated company in the industrial control system and information security field. It is also committed to being a research institute which plays a vital strategic position in the field of national scientific defense.

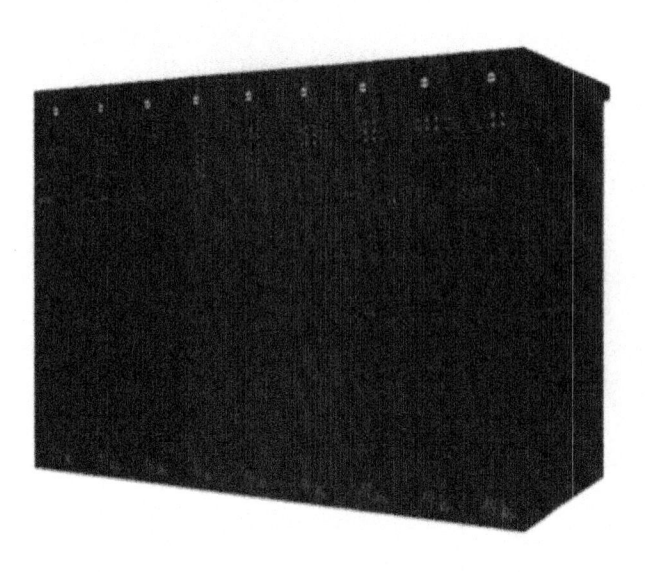

地址：北京市海淀区清华东路25号 邮编：100083
电话：010-66608989 邮箱：ncse@ncse.com.cn
Add: No. 25, Qinghua East Road, Haidian, Beijing, China Post Code: 100083
Tel: +8610-66608989 E-mail: ncse@ncse.com.cn

第六研究所

China Electronics Corp. (CEC). Booklet. The 6th Research Institute of China Electronics Corp. (1/2). 2016 Zhuhai Airshow. (51/54).

便携式电台综合测试仪
The portable radio integrated tester

便携式电台综合测试仪采用PXI总线架构，主要面向全军通信装备各级维修单位提供一种集音频信号发生及分析、射频信号发生、信号调解、频率测量、功率测量、电压测量等多种测试功能于一体的电台装备综合测试仪器，用于多种电台装备的现场原位级测试和维修，具有虚拟仪器、一键测试、流程测试、交互式诊断等智能化测试软件功能，能满足中长波、短波、超短波、多波段电台的测试和维修保障需求，实现军用电台的综合化、智能化、通用化测试，提高通信装备水平、战时技术保障。

The portable radio integrated tester adopts PXI bus architecture, which is mainly used to provide a set of audio signal generation and analysis, RF signal generation, signal demodulation, frequency measurement, power measurement, voltage measurement and other tests for all levels of military communication equipment maintenance units. Function testing and maintenance of radio equipment in the field, with a virtual instrument, a key test, process testing, interactive diagnostics and other intelligent test software function, it can meet the long wave, short wave, ultra-short wave, multi-band radio test and maintenance needs, to achieve military radio integrated, intelligent, universal test, to improve the level of communications equipment, technical support in wartime.

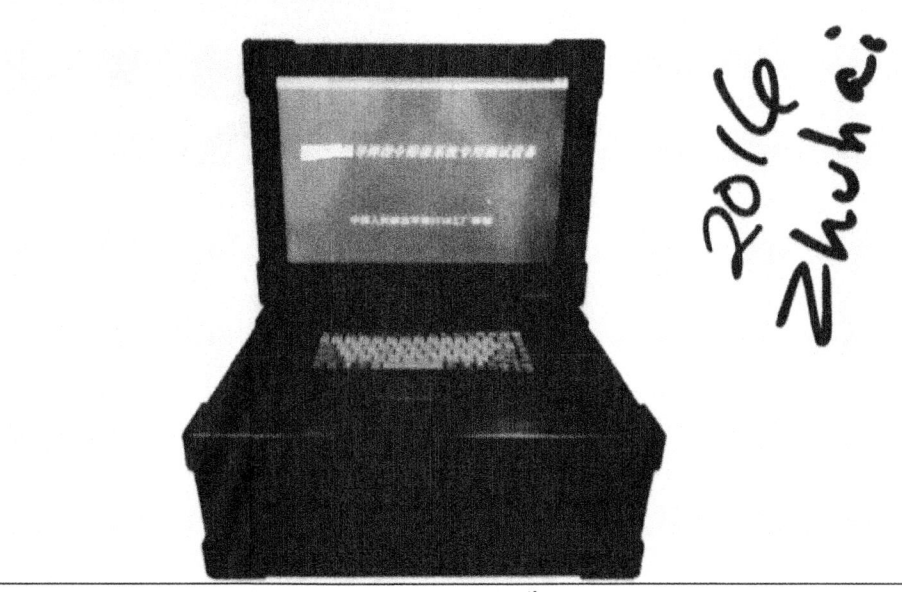

China Electronics Corp. (CEC). Booklet. The 6th Research Institute of China Electronics Corp. (2/2). The Portable Radio Integrated Tester. 2016 Zhuhai Airshow. (52/54).

中国电子进出口总公司
CEIEC

公司简介 Company Profile

中国电子进出口总公司（CEIEC）成立于1980年4月。经过多年的诚信经营，CEIEC已与全世界160多个国家地区建立了广泛的业务合作，为中国的改革开放和中国电子工业的发展做出了重要贡献。

CEIEC具有国际贸易、国际工程总承包、招标代理、展览广告等多种业务的甲级经营资质。2015年底，CEIEC总资产达298.91亿元人民币，当年实现销售收入375.3亿元人民币。

CEIEC的战略重点立足于打造防务系统集成、公共安全集成、海外工程集成、贸易服务集成四大主业。

CEIEC was founded in April, 1980. With years operation of honesty and diligence, CEIEC has built wide-ranged cooperation relationships with more than 160 countries and regions, and has made great contribution to the reforms and Opening-up of China and to the development of Chinese electronics industry.

CEIEC is honorably entitled to a number of A-grade certificates of world trade, international engineering, tendering, exhibition and advertisement. By the end of 2015, CEIEC's total assets and sales revenue has respectively reached RMB 29.89 billion and RMB 37.53billion.

At present, CEIEC's strategy is focusing on the following four "integrations": defense electronics system integration, public security Integration, overseas engineering integration, business solutions integration.

地址：北京市海淀区复兴路17号国海广场A座新中电大厦
电话：+86-10-52579999
传真：+86-10-52579000
邮编：100036
E-mail: celec@ceiec.com.cn
http://www.ceiec.com.cn

Add: New CEIEC Building, No. 17 Fuxing Road, Haidian District, Beijing
Tel: +86-10-52579999
Fax: +86-10-52579000
ZipCode: 100036
E-mail: ceiec@ceiec.com.cn
http://www.ceiec.com

China Electronics Corp. (CEC). Booklet. China National Electronics Import & Export Corp./CEIEC. (1/2). 2016 Zhuhai Airshow. (53/54).

公共安全一体化平台

2011年3月至2014年5月，中国电子进出口总公司承建了公共安全一体化平台，包含2个国家中心，5个区域中心，9个本地中心，3000个外设（包括摄像头、MDVR等），覆盖该国28万平方公里的1600万人口。

作为拉丁美洲最先进的一体化公共安全平台之一，该项目自建成以来，有效提高了该国政府的工作管理效率，经过三年的持续运行，该国的犯罪率下降了27%。在地震等灾害中，该平台在应急救援、资源协调等方面发挥了巨大的作用，受到了政府、民众和国际媒体的一致赞扬。

Integrated Public Security Platform

From 2011.3 to 2014.5, CEIEC built the Integrated Public Security Platform in some country in Latin America. The project contains 2 national command centers, 9 local command centers, 5 regional command centers and 3000 peripherals(including cameras, MDVR, etc.), covering this country 280,000 km² and 16,000,000 population.

As one of the most advanced integrated public security platform in Latin America, this system has supported the government to achieve great progress in their management efficiency. After 3 years' continuous operation, the crime rate in this country has been decreased to 27%. In the accident like earthquake, this system bears the test and plays an irreplaceable role in departments collaboration and emergency response which is highly praised by the government, people and international media.

China Electronics Corp. (CEC). Booklet. China National Electronics Import & Export Corp./CEIEC. (2/2). Integrated Public Security Platform. 2016 Zhuhai Airshow. (54/54).

Luoyang Institute of Electro-Optical Equipment; Aviation Industry Corporation of China (AVIC)

"LOONG EYE" LE120 Electro-optical Payload

Functions

- Detection: Visible or IR detection
- Searching: Manual, automatic searcning
- Stabilization: Gyro stabilization
- Tracking: Automatic tracking(extendable)
- Storage: Video/image storage

Specifications

- Dimension: ≤Φ120mm×220mm
- Weight: ≤2kg
- Rotation range: azimuth 360° continuous elevation −110°~+20°
- Stabilization accuracy: ≤1mrad
- Visible light sensor
  Color CCD: 1920×1080
  FOV: 5.4°×3°~50°×28°
  (Continuous zooming)
- IR sensor:
  Sensor: Uncooled 640×480 detector
  FOV: 9.3°×7.1°

EO Payload

Ground Control Bench

AVIATION INDUSTRY CORPORATION OF CHINA Luoyang Institute of Electro-optical Equipment
Tel:0379-63327181 Email: eoei@vip.sina.com

Loong Eye LE120 Electro-Optical Payload. Luoyang Institute of Electro-Optical Equipment. Aviation Industry Corp. of China (AVIC Optronics). Unknown Airshow (1/14).

"LOONG EYE" LE140 Electro-optical Payload

"AVIC Optronics"

Functions

- Detection: Visible or IR detection
- Searching: Manual, automatic searching
- Stabilization: Gyro stabilization
- HD imaging
- Storage: Video/image storage

EO Payload

Specifications

- Dimension: ≤Φ130mm×235mm
- Weight: ≤3.5kg
- Rotation range: Azimuth 360° continuous elevation −110°~+20°
- Stabilization accuracy: ≤1mrad
- Visible light sensor:
  Color CCD: Video definition 1920×1080
  Image definition: 24 Mega-pixel
- FOV: 26.4°×17.7°

Ground Control Bench (optional)

Loong Eye LE140 Electro-Optical Payload. Luoyang Institute of Electro-Optical Equipment. Aviation Industry Corp. of China (AVIC Optronics). Unknown Airshow (2/14).

"LOONG EYE" LE180 Electro-optical Payload

>> Functions

- Detection: Visible or IR detection and imaging, temperature measurement and pseudo-color output
- Searching: Manual and automatic searching
- Stabilization: Gyro-stabilized function
- Storage: Self-contained video, image storage or thermal image storage

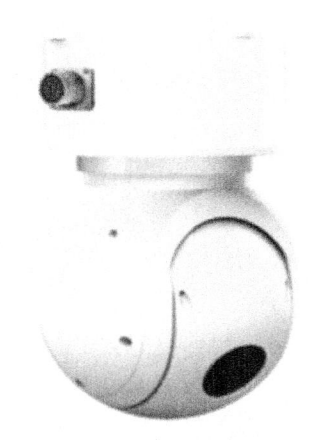

EO Payload

>> Specifications

- Dimension: ≤Φ180mm×330mm
- Weight: ≤5kg
- Rotation range: Azimuth 360° continuous, elevation −110°~+20°
- Stabilization accuracy: ≤1mrad
- Visible sensor
  Color CCD: 1920×1080
  FOV: 8.2°×4.6°~66°×38°
  (continuous zooming)
- IR sensor
  640×480 un-cooled IR detector
  FOV: 17°×13°

Ground Station (optional)

Loong Eye LE180 Electro-Optical Payload. Luoyang Institute of Electro-Optical Equipment. Aviation Industry Corp. of China (AVIC Optronics). Unknown Airshow (3/14).

"LOONG EYE" LE220 Electro-optical Payload

Functions

- Detection: Take high-definition video and pictures
- Searching: Manual and automatic searching, etc
- Stabilization: Gyro-stabilized function
- Storage: Self-contained video, image storage

EO Payload

Specifications

- Dimension: ≤Φ220mm×350mm
- Weight: ≤6kg
- Rotation range: Azimuth 360° continuous elevation −110°~+20°
- Stabilization accuracy: ≤0.2mrad
- Visible sensor:
  Color CCD: 1920×1080
  FOV: 4.8°×2.6°~64°×36°
  (continuous zooming)
- Visible camera(picture)
  Resolution of camera: 24 mega-pixel
  FOV of camera: 14.9°×10°
  Focal length: 135mm

Ground Station (Optional)

Loong Eye LE220 Electro-Optical Payload. Luoyang Institute of Electro-Optical Equipment. Aviation Industry Corp. of China (AVIC Optronics). Unknown Airshow (4/14).

"LOONG EYE" LE260 EO Payload

Functions

- To search, identify and track targets in adverse weather and in day/night
- To implement laser-ranging for targets
- To output video signals of outside sceneries

Specifications

- Cooled MWIR thermal imager 24°×18°/2°×1.5° (fixed FOV)
- Color/black&white TV sight, 24°×18°/2°×1.5° (continuous zooming)
- 1.54 μm laser range-finder, ranging scope: 300m~8000m
- Rotation range: Azimuth 360° continuous elevation −120°~+30° (adjustable upon request)
- Stabilization accuracy: ≤80 μrad
- Weight: ≤18kg
- Dimension: ≤Φ260mm×398mm

Configuration

- LE260 I : IR thermal imager + TV sight
- LE260 II : IR thermal imager + laser range finder
- LE260 III : TV sight + laser range finder

AVIATION INDUSTRY CORPORATION OF CHINA Luoyang Institute of Electro-optical Equipment
Tel:0379-63327181   Email: eoei@vip.sina.com

EO Turret

Elevating Mechanism (Optional)

Loong Eye LE260 Electro-Optical Payload. Luoyang Institute of Electro-Optical Equipment. Aviation Industry Corp. of China (AVIC Optronics). Unknown Airshow (5/14).

"LOONG EYE" 330D Electro-optical Payload

It could provide stable and clear IR & visible images, realize such functions as searching, tracking, laser ranging and positioning designated areas day and night; it is applicable to various kinds of manned and unmanned aerial platforms.

Main Functions

- Detection: IR & visible detection imaging
- Searching: Manual and automatic searching modes
- Positioning: Passive target positioning
- Tracking: Automatic tracking and geographical tracking
- Fog penetration: Image enhancement fog penetration

Payload

Specifications

- Weight: ≤28kg
- Size: ≤Φ330mm×470mm
- Rotation range: 360°continuous in azimuth, -120°~ +40° in elevation
- Stabilization accuracy: ≤25 μrad
- IR
  IR detector: Cooled 640×512 MWIR
  FOV: Three FOVs, 4°×3°, 10°×7.5°, 24°×18°
- Visible light
  Color CCD, 1920×1080
  FOV: 3.7°×2.1°~ 36.5°×21°
- Positioning accuracy: At the distance of 10km, the real-time positioning accuracy shall be no more than 15m

Loong Eye LE330D Electro-Optical Payload. Luoyang Institute of Electro-Optical Equipment. Aviation Industry Corp. of China (AVIC Optronics). Unknown Airshow (6/14).

'LOONG EYE' LE350 EO Payload

Functions

- To search, identify, track and assistance navigation scenery video/image output in adverse weather and in day/night
- To perform laser ranging & designation and guide laser-guided weapons
- Record data and video information (optional)

Specifications

- Cooled MWIR IR thermal imager
  WFOV 24°×18°
  NFOV 1.2°×0.9° (fixed FOV)
- Color/B/W TV Sight, 24°×18°~ 1.2°×0.9° (continuous adjustment)
- Digital Camera, Pixel number: ≥4k×5k
- 1.064 μm laser range-finder, ranging scope: 300m~15000m
- Laser designator with output energy above 50mJ
  divergence angle: ≤0.5mrad
- Rotation range: Azimuth 360° continuous
  elevation −120°~+30° (adjustable upon request)
- Stabilization accuracy: ≤40 μrad
- Outline dimension: ≤Φ350mm×500mm, weight: ≤32kg

EO Turret

Control Handle

Recorder (Optional)

Sensor Configuration

- LE350 I : IR thermal imager + TV sight
- LE350 II : IR thermal imager + laser designator
- LE350 III : IR thermal imager + laser range finder
- LE350 IV : IR thermal imager + TV sight + laser range finder
- LE350 V : IR thermal imager + TV sight + laser designator
- LE350 VI : TV sight + laser range finder
- LE350 VII : TV sight + laser designator
- LE350 VIII : IR thermal imager + TV sight + CCD digital camera

Loong Eye LE350 Electro-Optical Payload. Luoyang Institute of Electro-Optical Equipment. Aviation Industry Corp. of China (AVIC Optronics). Unknown Airshow (7/14).

'LOONG EYE' LE380 EO Payload

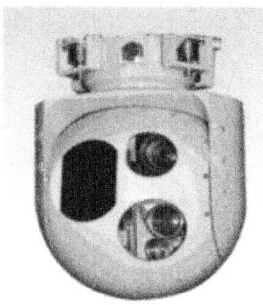

Functions

- To search, identify, track, aim at targets and assist navigation in adverse weather and under day/night, output scene video or photos, capable of recording data and video information
- To perform laser ranging & designation, and guide laser-guided weapons
- Target geo-location and multi-target detection cueing
- Scanning speed and range can be set for auto-scanning
- Picture-in-picture display

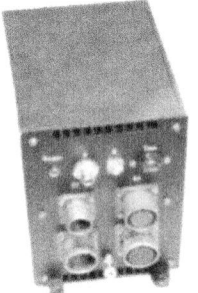

EO Turret

Electronic Unit

Specifications

- Cooled MWIR/LWIR thermal imager, 24°×18° / 1.2°×0.9° (fixed or continuous zooming)
- Color/ black & white TV Sight, 24°×18°/ 0.8°×0.6° (fixed or continuous zooming)
- Digital camera with pixel above 4k×5k
- Laser range finder with wavelength of 1.064μm and ranging scope of 300m~15000m
- Laser designator with output energy above 80mJ and divergence angle no more than 0.35mrad
- Rotation range: azimuth 360° continuous elevation −130°~+30° (adjustable upon request)
- Stabilization accuracy: ≤30μrad
- Dimension does not exceed Φ380mm×555mm and weight is no more than 39kg

Sensor Configuration

- LE380 I : IR thermal imager + TV sight
- LE380 II : IR thermal imager + laser designator
- LE380 III: IR thermal imager + laser range finder
- LE380 IV : IR thermal imager + TV sight + laser range finder
- LE380 V : IR thermal imager + TV sight + laser designator
- LE380 VI: TV sight + laser range finder
- LE380 VII: TV sight + laser designator
- LE380 VIII: IR thermal imager + TV sight + CCD digital camera

Loong Eye LE380 Electro-Optical Payload. Luoyang Institute of Electro-Optical Equipment. Aviation Industry Corp. of China (AVIC Optronics). Unknown Airshow (8/14).

"LOONG EYE" LE400 Electro-optical Payload

Functions

- To search, identify, track, aim at targets and assist navigation in adverse weather and under day/night, output scene video or photos, capable of recording data and video information
- To perform laser ranging & designation, and guide laser-guided weapons
- Picture-in-picture display

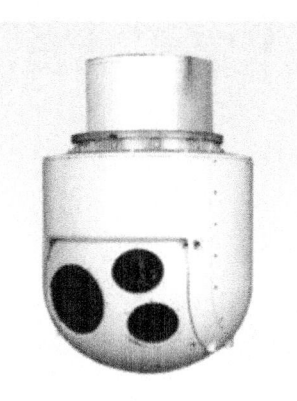

EO Turret

Control Handle

Specifications

- Cooled MWIR/LWIR thermal imager, 24°×18° / 1.2°×0.9° (fixed or continuous zooming)
- Color/ black & white TV sight, 24°×18°/ 0.8°×0.6° (fixed or continuous zooming)
- Digital camera with pixel above 4k×5k
- Laser range finder with wavelength of 1.54μm and ranging scope of 300m~20000m
- Laser designator with output energy above 80mJ and divergence angle no more than 0.3mrad
- Rotation range: Azimuth 360° continuous; elevation −120°~+30° (adjustable upon user's request)
- Stabilization accuracy: ≤30μrad
- Dimension does not exceed Φ400mm×700mm, and weight is no more than 55kg

Electronic Unit

Sensor Configuration

- LE400 I : IR thermal imager + TV sight + laser range finder
- LE400 II : IR thermal imager + TV sight + laser designator
- LE400III: IR thermal imager + TV sight + CCD digital camera
- LE400IV: IR thermal imager + TV sight + CCD digital camera + laser range finder

Loong Eye LE400 Electro-Optical Payload. Luoyang Institute of Electro-Optical Equipment. Aviation Industry Corp. of China (AVIC Optronics). Unknown Airshow (9/14).

"LOONG EYE" LE430 Electro-optical Payload

Functions

- To search, identify, track and aim at targets, as well as assist navigation day and night and in adverse weather, output scene video or photo
- To perform laser ranging & designation, and guide laser-guided weapons
- To implement calibration for three optical axes of IR, TV and laser

EO Turret

Control Handle

Electronic Unit (Optional)

Specifications

- Cooled MWIR/LWIR thermal imager, 24°×18°/0.8°×0.6° (fixed or continuous zooming);
- Color/ black & white TV sight, 24°×18°/0.4°×0.3° (fixed or continuous zooming)
- Digital camera with pixel above 4k×5k
- Laser range finder with wavelength of 1.064μm and ranging scope of 300m~20000m
- Laser designator with output energy of more than 80mJ and divergence angle of no more than 0.3mrad
- Rotation range: 360° continuous in azimuth, -120°~+30° in elevation (adjustable upon request)
- Stabilization accuracy: ≤30μrad
- Dimension does not exceed Φ430mm×700mm, and weight is no more than 55kg

Sensor Configuration

- LE430 I : IR thermal imager + TV sight + Laser range finer
- LE430 II : IR thermal imager + TV sight + laser designator
- LE430 III: IR thermal imager + TV sight + CCD digital camera
- LE430 IV: IR thermal imager + TV sight + CCD digital camera + laser range finder

Loong Eye LE430 Electro-Optical Payload. Luoyang Institute of Electro-Optical Equipment. Aviation Industry Corp. of China (AVIC Optronics). Unknown Airshow (10/14).

'LOONG EYE' LE450 EO Payload

Functions

- To search, identify, track and aim at targets, as well as assist navigation day and night and in adverse weather, output scene video or photo
- With LOS stabilization and (manual, automatic) tracking functions, and be able to implement calibration for three optical axes
- To perform laser ranging & designation, and guide laser-guided weapons

EO Turret

Control Handle

Specifications

- Cooled MWIR/LWIR thermal imager, 24°×18°/0.8°×0.6° (fixed or continuous zooming)
- Color/ black and white TV sight, 24°×18°/0.4°×0.3° (fixed or continuous zooming)
- Digital camera with pixel above 4k×5k
- Laser range finder with wavelength of 1.064μm and ranging scope of 300m~20000m
- Laser designator with output energy more than 100mJ and divergence angle no more than 0.3mrad
- Rotation range: 360° continuous in azimuth; -120°~+30° in elevation (adjustable upon request)
- Stabilization accuracy: ≤30 μrad
- Dimension does not exceed Φ450mm×700mm, and weight is no more than 57kg

Sensor Configuration

- LE450 I : IR thermal imager + TV sight + laser range finder
- LE450 II : IR thermal imager + TV sight + laser designator
- LE450 III : IR thermal imager + TV sight + CCD digital camera + laser designator

Loong Eye LE450 Electro-Optical Payload. Luoyang Institute of Electro-Optical Equipment. Aviation Industry Corp. of China (AVIC Optronics). Unknown Airshow (11/14).

'LOONG EYE' 450H EO Payload

"LOONG EYE" 450H EO Payload may provide stable and clear IR image, realize searching of designated area as well as targeting, tracking, laser ranging and positioning of targets day and night; it is mainly used for target searching, recognizing, identification, tracking and positioning, and is applicable to various kinds of manned/unmanned aerial vehicle and shipboard platform.

Main Features

- Large area array IR imaging
- Target positioning of high-accuracy
- 1080P high-definition visible imaging
- High-power laser
- Automatic tracking of target
- Image enhancement

Specifications

- Weight: ≤60kg
- size: ≤Φ450mm×700mm

Movement range

- Rotation range: 360° continuous in azimuth, -120°~ +30° in elevation
- Stabilization accuracy: ≤25 μrad
- IR
  IR detector: cooled 640×512 MWIR
  FOV: 1.3°×1.05°~ 25.8°×20.64°
- Visible light
  Color CCD, 1920×1080
  FOV: 1.2°×0.675°~ 24°×13.5°
- Laser rangefinder
  Wavelength: 1.064 μm
  Ranging distance: ≥20km
  Ranging accuracy: 5m
- Positioning accuracy: At the distance of 10km, the real-time positioning accuracy shall be no more than 15m

EO payload

AVIATION INDUSTRY CORPORATION OF CHINA Luoyang Institute of Electro-optical Equipment
Tel:0379-63327181   Email: eoei@vip.sina.com

Loong Eye LE450H Electro-Optical Payload. Luoyang Institute of Electro-Optical Equipment. Aviation Industry Corp. of China (AVIC Optronics). Unknown Airshow (12/14).

'LOONG EYE' LE480 Model EO Turret

Functions

- Search, identify, track and aim at targets in adverse weather and in day/night
- Implement reconnaissance and taking pictures in day
- Output LOS information and video signals of outside sceneries
- Laser range & designate targets and guide laser-guided weapons;
- Picture-in-picture display
- Implement aided navigation in adverse weather
- Implement 3-axis calibrations for IR, TV and laser optical axis

Specification

- IR Thermal Imager
  Cooled MWIR/LWIR detector
  FOV: max.FOV of 24°×18°; min. FOV of 0.8°×0.6° (fixed or continuous zooming)
- TV Sight
  Color/ monochrome CCD
  FOV: max. FOV of 24°×18°; min. FOV of 0.4°×0.3° (fixed or continuous zooming)
- Digital Camera
  Pixel number: ≥4k×5k; Focal length: ≥110mm
- Laser Range Finder
  Wavelength: 1.064μm; Ranging scope: 300m~20000m
- Laser Designator
  Energy: ≥100mJ; Divergence angle: ≤0.3mrad
  Designation frequency: 20Hz (adjustable)
- Rotation range: azimuth 360° continuous; elevation −120°~+30° (adjustable upon user's request)
- Stabilization accuracy: ≤30μrad
- Weight: ≤80kg
- Dimension: ≤Φ480mm×730mm

Sensor Configuration

- LE480 I model: IR thermal imager + TV sight + laser range finder
- LE480 II model: IR thermal imager + TV sight + laser designator
- LE480 III model: IR thermal imager + TV sight + CCD digital camera + laser designator

Loong Eye LE480 Electro-Optical Payload. Luoyang Institute of Electro-Optical Equipment. Aviation Industry Corp. of China (AVIC Optronics). Unknown Airshow (13/14).

'LOONG EYE' LE500 Model EO Turret

Functions

- Search, identify, track and aim at targets in adverse weather and in day/nigh
- Implement reconnaissance and taking pictures in day
- Contrast tracking and shape tracking (optional)
- Output LOS information and video information of outside sceneries
- Laser range & designate targets and guide laser-guided weapons
- Picture-in-picture display
- Moving target automatic identification and cueing
- Implement aided navigation in adverse weather
- Implement 3-axis calibrations for IR, TV and laser optical axis

EO Turret

Specification

- IR Thermal Imager
  Cooled MWIR/LWIR detector
  FOV: max. FOV of 24°×18°; min. FOV of 0.8°×0.6° (fixed or continuous zooming)
- TV Sight
  Color/ monochrome CCD
  FOV: max. FOV of 24°×18°; min. FOV of 0.4°×0.3° (fixed or continuous zooming)
- Digital Camera
  Pixel number: ≥4k×5k
  Focal length: ≥110mm
- Laser Designator
  Wavelength: 1.064μm
  Ranging scope: 300m~20000m
  Energy: ≥100mJ
  Divergence angle: ≤0.25mrad
  Designation frequency: 20Hz (adjustable)
- Rotation range: azimuth 360° continuous;
  elevation −120°~+30° (adjustable upon user's request)
- Stabilization accuracy: ≤30μrad
- Weight: ≤70kg
- Dimension: ≤Φ500mm×740mm

Sensor Configuration

- LE500−I model: IR thermal imager + TV sight + laser designator
- LE500−II model: IR thermal imager + TV sight + CCD digital camera + laser designator

AVIATION INDUSTRY CORPORATION OF CHINA Luoyang Institute of Electro-optical Equipment
Tel:0379-63327181   Email: eoei@vip.sina.com

Loong Eye LE500 Electro-Optical Payload. Luoyang Institute of Electro-Optical Equipment. Aviation Industry Corp. of China (AVIC Optronics). Unknown Airshow (14/14).

INDEX

2-Frequency Shift Keying (2-FSK) **Volume 2:** 212
20W/125W HF Digital Frequency-Hopping Radio **Volume 2:** 209
6U CPC1 Card **Volume 1:** 86-99
2K resolution **Volume 1:** 86-88
6th Research Institute of China Electronics Corp. (*see also* China Electronics
    Corp./CEC) **Volume 1:** 102-103
16QAM **Volume 2:** 259-260
1:N matching accuracy segment (facial recognition) **Volume 1:** 21
1:N comparison **Volume 1:** 23
27th Research Institute of CETC **Volume 2:** 240-241, 253-256
3A hospitals **Volume 1:** 20, 26, 32, 34
91B Radar **Volume 1:** 73
3-ring antenna **Volume 2:** 206
137C radioactive source **Volume 1:** 76
3-Level Architecture **Volume 2:** 250
3D **Volume 1:** 55-56, 73, 83-84, 86-88; **Volume 2:** 168, 184
722 Factory **Volume 1:** 65-68
714 Weather Radar **Volume 1:** 72
863 Program (State High-Tech Development Plan) **Volume 1:** 94
1553B **Volume 2:** 270-272
9350 antenna **Volume 2:** 203

A

A Grade Certification **Volume 1:** 104
AAL2 adaptation **Volume 2:** 216
Access and Switching and Relay Vehicle **Volume 2:** 280-281
Access Controller Access (ACA) **Volume 1:** 28
accurate target attack **Volume 2:** 250
ACM International Collegiate Programming Contest (ACM-ICPC) **Volume 1:** 29
Acoustic Denial Vehicle **Volume 2:** 171-172
acoustic monitoring system **Volume 2:** 170
active electronic counter measures **Volume 2:** 180-181
ad hoc network **Volume 2:** 192-195
adaptative anti-radar shielding algorithm **Volume 1:** 55
adaptive frequency change **Volume 2:** 247-248
Advanced Data Terminal (ADT) **Volume 2:** 278
advanced SMT processing **Volume 1:** 55
Aeronautical Radio, Inc. **Volume 1:** 83, 85
AGC cheating **Volume 2:** 182

Ah SLA battery **Volume 2:** 202
AICARE Range **Volume 1:** 25-26
air communication network **Volume 2:** 250-251, 279
air cooling structure **Volume 1:** 75
air defense **Volume 1:** 1, 53, 55-56, 62, 73; **Volume 2:** 160, 162-164, 183-185, 215, 220, 228, 285
Air Force **Volume 1:** 53, 55, 60, 79, 81-84; **Volume 2:** 225, 282
air processing channels **Volume 1:** 56
air search **Volume 1:** 56
air traffic control **Volume 2:** 190
air-to-air link **Volume 2:** 192
air-to-ground data link **Volume 2:** 192
air-to-ground missile guidance **Volume 2:** 219
Airborne Early Warning and Control Aircraft (AEW&C) **Volume 2:** 215
airborne payload equipment **Volume 2:** 213
airborne radar **Volume 2:** 182
airborne terminal **Volume 2:** 192
airborne transceiver **Volume 2:** 194
aircraft **Volume 1:** 60, 79; **Volume 2:** 160, 186, 191-192, 215, 254
aircraft carrier **Volume 1:** 79
airports **Volume 1:** 22, 23, 28, 46, 68, 72; **Volume 2:** 165, 190
Alan Trusted Operating System **Volume 1:** 78-79
alarm delay **Volume 1:** 2, 15, 18V
Algebraic Code Excited Linear Prediction (ACELP) **Volume 2:** 216
algorithm **Volume 1:** 2, 20-21, 23, 55, 95 ; **Volume 2:** 261-262, 264-265
Alibaba **Volume 1:** 29, 31
alternate routing **Volume 2:** 280
AM **Volume 2:** 238, 263
AMDE4690/E6465 **Volume 1:** 86-99
AN-W1 AN-Watch **Volume 1:** 79, 81
Analog Voice CVSD **Volume 2:** 247, 249
angel funding **Volume 1:** 32, 35
anode-controlled electron gun **Volume 1:** 77
Anovo-SatTour 300A **Volume 2:** 196-197
antenna **Volume 1:** 54-55; **Volume 2:** 176-177, 180-182, 184, 188-189, 191-192, 194, 197-198, 203, 205-206, 209-210, 213, 225-226, 228, 247, 259-260, 268, 273, 275-277, 282-283, 285
anti-acoustic **Volume 1:** 79
anti-acquisition **Volume 2:** 228
Anti-Chemical Vehicle **Volume 1:** 62; **Volume 2:** 271
Anti-Chemical Warfare **Volume 1:** 62
anti-drip enclosure protection **Volume 2:** 277

anti-drug trafficking **Volume 2:** 239
anti-dust **Volume 1:** 79, 81
anti-error-bit ability **Volume 2:** 221
anti-ESM **Volume 2:** 186
anti-fading **Volume 1:** 50; **Volume 2:** 221-222, 228, 230, 243, 284
anti-fading BITE (Built-In Test Equipment) Function **Volume 1:** 50
anti-interception **Volume 2:** 193, 228
anti-jamming **Volume 1:** 55-56, 73; **Volume 2:** 184, 186, 192-193, 209, 225, 228-229,
    257, 259, 278, 282, 285 (see also jamming)
anti-jolting performance **Volume 2:** 189
anti-multi-path **Volume 2:** 257
anti-overlap function **Volume 2:** 189
anti-piracy **Volume 1:** 68
anti-rapid-fading capability **Volume 2:** 225, 282
anti-rotor shielding algorithm **Volume 1:** 54
anti-seismic **Volume 1:** 76; **Volume 2:** 277
anti-stealth **Volume 1:** 73
anti-tank weapon **Volume 2:** 169, 273-274
anti-terrorism **Volume 1:** 1, 48, 68; **Volume 2:** 160, 169, 172, 239
Anti-Terrorism Mission Vehicle **Volume 1:** 48
Anti-UAV **Volume 1:** 41-43, 48
anti-virus software **Volume 2:** 234-235
anti-wiretapping **Volume 2:** 208
Apsara Stack **Volume 1:** 29, 31
ARINC 429 (Aeronautical Radio, Inc.) **Volume 1:** 83, 85
ARM CPU Eco-System **Volume 1:** 93, 98
ARM64 **Volume 1:** 93
ARM9101 Airborne Radar Warning Receiver **Volume 1:** 60
armored vehicle **Volume 2:** 170-172, 266
Army Tactical Command and Control System (ATCCS) **Volume 2:** 214
Army Tactical Communication Network **Volume 2:** 250-251, 279
artificial intelligence (AI) **Volume 1:** 2, 3, 20-23, 26-27, 29-30, 32-33, 36
artillery **Volume 1:** 58, 62, 64; **Volume 2:** 186, 215, 241
Asia Square Tower **Volume 1:** 39
assistant navigation **Volume 1:** 56
Association for Computing Machinery (ACM) **Volume 1:** 29
asynchronization **Volume 2:** 185, 222, 243, 284
asynchronized jamming resistance **Volume 2:** 185
asynchronous semi-duplex bus transmission mode **Volume 1:** 64
ATM **Volume 1:** 27-28; **Volume 2:** 214, 216, 245-246, 280
ATM adaptation layer 2 (AAL2) **Volume 2:** 216
ATM broadcasting **Volume 2:** 214

ATM packet switching **Volume 2:** 245
atoll parabolic **Volume 2:** 176
atomic clock **Volume 2:** 230, 232, 285
ATR Cabinet **Volume 1:** 54
attack control **Volume 1:** 91-92
attack decision-making chain **Volume 1:** 66
Audio Bus Architecture. **Volume 1:** 103
Audio Frequency (AF) **Volume 2:** 212
audio response **Volume 2:** 204
authentication and authorization system **Volume 2:** 234-235
AUTO Negotiation **Volume 1:** 89
auto-failure-location **Volume 2:** 216
autoimmunization **Volume 2:** 282
automated key distribution **Volume 2:** 261
Automatic Gain Control (AGC) **Volume 2:** 212
automatic identification of in-band data and G3 facsimile **Volume 2:** 216
automatic power control **Volume 2:** 247-248
Automatic Repeat Request (ARQ) **Volume 2:** 264
automatic tracking **Volume 1:** 107
auxiliary intelligence collection **Volume 2:** 220
Auxiliary Police SDN BHD (Singapore) **Volume 1:** 36-38, 40
aviation electronics **Volume 1:** 1; **Volume 2:** 160
Aviation Industry Corporation of China (AVIC) **Volume 1:** 106-120
azimuth measurement **Volume 2:** 201
AZURE Cloud Computing Platform (Microsoft) **Volume 1:** 3, 32-33

B

B.Tx/A.Rx **Volume 2:** 228
backbone trunk communication **Volume 2:** 225, 243, 247-248, 259, 282
Balloon Antenna Anchor Vehicle **Volume 1:** 69-70
banking **Volume 1:** 27-28 32, 35-38, 40, 98
Banyan Capital **Volume 1:** 3, 20, 32, 35
barrage-jarring **Volume 2:** 237
beacon receiver **Volume 2:** 198
bearing rods **Volume 2:** 268, 272
Beidou satellite system **Volume 1:** 77-79; **Volume 2:** 177
Beijing Exhibition Center **Volume 1:** 1; **Volume 2:** 160
Beijing Shenofeifan Electronic System Technology Development Co. Ltd. **Volume 1:** 69-70
Beijing Xinlong Electronics New Technology Co. Ltd. **Volume 1:** 1; **Volume 2:** 160
Beyond Line of Sight (BLOS) **Volume 1:** 50; **Volume 2:** 225, 228, 282, 284-285

Bialkali **Volume 1:** 76
binocular camera **Volume 1:** 28
bio-medicine **Volume 1:** 95
biochemical **Volume 1:** 62
Bit Error Rate (BER) **Volume 2:** 216, 225, 228, 247-248, 282, 285
BITE (built-in test equipment) Function **Volume 1:** 50
BJ48 Waveguide Vehicle **Volume 1**: 50
blacklist **Volume 1:** 2, 12, 15, 18, 24
blimp/airship/balloon **Volume 1:** 69-70
blind zone **Volume 2:** 195
Beyond Line of Sight (BLOS) **Volume 1:** 50
BM-2021F **Volume 1:** 75
Boao Forum **Volume 1:** 2, 20, 34
body-worn camera **Volume 1:** 2
bone age measurement **Volume 1:** 25
border security **Volume 2:** 190
brain science **Volume 1:** 29-30
Brazil **Volume 1:** 2, 20
BRICS Summit (Brazil, Russia, India, China, South Africa) **Volume 1:** 2, 20
broadband **Volume 1:** 78-79; **Volume 2:** 185, 192-193, 250-251, 280, 284
broadband active jamming **Volume 2:** 185
broadband optical transmission network **Volume 2:** 250-251
broadly plateau characteristic **Volume 1:** 76
bUILd **Volume 1:** 36-38, 40
built-in secondary-level multiplexer **Volume 2:** 259
built-in test equipment **Volume 1:** 50
bulk information **Volume 2:** 282
burst-point correction range **Volume 2:** 241

C

C interface **Volume 2:** 243-244
C-band **Volume 1:** 56, 75; **Volume 2:** 183, 213, 220
camera **Volume 1:** 2, 28, 78, 114, 116; **Volume 2:** 166
campus network **Volume 1:** 91-92
campus security **Volume 1:** 28
CAN standard interface **Volume 2:** 270
cancer **Volume 1:** 36, 38
Canon Group **Volume 1:** 36-38, 40
Capability Maturity Model Levels (CMMI) **Volume 1:** 98
car brand recognition system **Volume 1:** 23
car location tracking **Volume 1:** 23

car trunk **Volume 2:** 197
carbon fiber **Volume 2:** 176, 198
cardless ATM withdrawals **Volume 1:** 27-28
casino **Volume 1:** 24
Cathode-Ray Tube (CRT) **Volume 1:** 78
CCD Digital Camera **Volume 1:** 114, 116-117, 120; **Volume 2:** 241
CCIR5120-4 **Volume 2:** 176
CCMA-3000 HF/VHF/UHF surveillance and analysis receiver **Volume 2:** 212
Central America **Volume 1:** 104-105
Central Processing Unit (CPU) **Volume 1:** 86-99
centrex function **Volume 2:** 236
centroid **Volume 2:** 258
ceramics **Volume 1:** 74-75, 77
certificate application **Volume 1:** 79
Certis **Volume 1:** 36-38, 40
chaff **Volume 2:** 252, 258
Changhai Airport Group **Volume 1:** 36-38, 40
Changjiang Electronics Group Co. Ltd. **Volume 1:** 53-56
Changsha Xiangji-Haidun Technology Co. Ltd. **Volume 1:** 82-89
chemical **Volume 1:** 62
Chengdu TimesTech Co. Ltd. **Volume 2:** 173-178
chest CT scan **Volume 1:** 32
China Aerospace Long-March International (ALIT) **Volume 2:** 187-198
China Border Patrol **Volume 1:** 36-38, 40
China Customs **Volume 1:** 36-38, 40
China Defence Electronics Application Forum **Volume 1:** 1; **Volume 2:** 160
China Electronic International Exhibition and Ads **Volume 2:** 160
China Electronics Corporation (CEC/6th Research Institute of China Electronics Corp.)
    Volume 1: 1, 51-107; **Volume 2:** 160
China Electronics International Exhibition & Advertising Co. Ltd. (CEIEC) **Volume 1:** 1
China Electronics Technology Group Corp. (CETC) **Volume 1:** 1; **Volume 2:** 211-216,
    221-286
China Immigration Inspection **Volume 1:** 36-38, 40
China International Defence Electronics Exhibit (CIDEX) **Volume 1:** 1; **Volume 2:** 160
China Linux OS Market **Volume 1:** 98
China Merchant Bank **Volume 1:** 27, 32, 35-38, 40
China National Electronic Import and Export Corp (CEIEC) **Volume 1:** 1, 104-105;
    Volume 2: 160
China National Precision Machinery Import and Export Corp. (CPMIEC) **Volume 2:**
161-170
China North Industries Group Corp. (NORINCO) **Volume 2:** 199-210

China Shipbuilding and Offshore International Co. (CSOC) **Volume 2:** 179-186
China Standard Software Co. Ltd. (CS2C/S2C) **Volume 1:** 97
China Unionpay **Volume 1:** 36-38, 40
Chinese installation interface **Volume 1:** 94
Chinese language characters **Volume 1:** 83-84
Chubb (United Technologies) **Volume 1:** 36-38, 40
ciphered voice **Volume 2:** 216
circuit status monitor **Volume 2:** 225, 282
Cisco Private Ltd. **Volume 1:** 36-38, 40
civil anti-interception communication system **Volume 2:** 257
clinical intelligent research platform **Volume 1:** 26
clock synchronization scheme **Volume 2:** 222, 236
Close-In Weapon System (CIWS) **Volume 2:** 183, 186
cloud **Volume 1:** 29, 31, 93, 95, 96
cluster malfunction **Volume 1:** 99
CM15 Certification **Volume 1:** 7
CMMI-5 (Capability Maturity Model Levels/Level 5) **Volume 1:** 98
coastal defense **Volume 2:** 190
coastal monitoring **Volume 2:** 169
coastal radar **Volume 1:** 66
Code Division Multiple Access (CDMA) **Volume 2:** 208, 257
codec **Volume 2:** 222, 243
coded orthogonal frequency-division multiplexing (COFDM) technology **Volume 2:** 194
combat type **Volume 1:** 2, 17
combox **Volume 2:** 163
Command and Control (C2) **Volume 1:** 48
command and control software **Volume 2:** 163
Command Vehicle **Volume 2:** 271
Command, Control, Communications, Computers, Intelligence (C4I) **Volume 2:** 286
Command, Control, Communications, Intelligence (C3I) **Volume 2:** 225, 252, 282
commercial detonation **Volume 2:** 195
communication and safety inspection equipment **Volume 2:** 200-210
communication encryption **Volume 1:** 1; **Volume 2:** 160
Communication Intelligence (COMINT) **Volume 2:** 238
communication relay **Volume 2:** 278, 284
Communication Vehicle **Volume 2:** 271
Communications-On-The-Move (COTM) Satellite Communication Vehicle **Volume 2:** 176
compass **Volume 2:** 177
Comprehensive Control Vehicle **Volume 1:** 69-70
computational photography **Volume 1:** 29-30
concentrated monitoring function **Volume 2:** 284

Conjugate-Structure – Algebraic Code Excited Linear Prediction (CS-ACELP) **Volume 2:** 216
Constant False Alarm Rate (CFAR) **Volume 2:** 184
Continuous Wave (CW) **Volume 1:** 77; **Volume 2:** 164, 233
Continuously Variable Slope Delta (CVSD) **Volume 2:** 221, 247, 249
Convolution Code RS **Volume 2:** 257
Corad Electronic Equipment Co., Ltd. **Volume 1:** 57-60
core switch **Volume 1:** 91-92
Courant Institute of Mathematical Sciences at New York University (NYU) **Volume 1:** 29-30
CPU: Freescale Power PC P1022/T1024 **Volume 1:** 86-99
credit investigation **Volume 1:** 28
crime **Volume 1:** 2, 105 (*see also* police)
critical infrastructure **Volume 1:** 22
cross polarization **Volume 2:** 198
cross-platform **Volume 1:** 101
cruise missile **Volume 2:** 183
cryptographic algorithms **Volume 1:** 95
cryptography **Volume 2:** 270, 234, 273, 276-277
crystal element **Volume 1:** 78-79
CS&S Informatica System Engineering Co. Ltd. **Volume 1:** 90
CT scan **Volume 1:** 32, 34
CT-300E/CT-400E TT&C System **Volume 2:** 212
customized algorithm **Volume 2:** 264-265
Customs Office **Volume 1:** 36, 38
cyber security **Volume 1:** 3, 69-70
cyberspace **Volume 2:** 234
Czech Republic **Volume 1:** 1; **Volume 2:** 160

D

D Interface **Volume 2:** 247, 249
DA-6 Tactical Internet Controller (TIC) **Volume 2:** 214-215
Dali, India **Volume 1:** 72
data annotation **Volume 1:** 26
data auxiliary channel **Volume 2:** 247-248, 259
data center aggregation **Volume 1:** 91-92, 96
data channel encryption **Volume 1:** 50
data communication network **Volume 2:** 169
data encryption **Volume 2:** 246
data ETL (extract transform and load) **Volume 1:** 26
data exchange **Volume 2:** 271, 274

data fusion **Volume 1:** 55-56
data interaction implementation **Volume 2:** 222
data modulation mode **Volume 1:** 62-63
data processing **Volume 1:** 95
data transmission rate **Volume 1:** 62; **Volume 2:** 209
dB SINAD **Volume 2:** 202
DDR4 **Volume 1:** 2
DDS **Volume 1:** 46
dead zone pairing **Volume 2:** 189
deception jamming **Volume 2:** 220
deep fading **Volume 2:** 216
deep learning **Volume 1:** 25-26, 29-30
Default Integrated Sogou Input Method Support WPS **Volume 1:** 94
Defence Services Asia (DSA/Malaysia) **Volume 1:** 2-10, 15-21, 23-29, 32-38, 40-50;
    Volume 2: 162-170, 179-186, 194-210, 217-219, 227-228, 240-241, 253-254
defend enclave boundary **Volume 2:** 234-235
defend the computing environment **Volume 2:** 234-235
Defense Data Network (DDN) **Volume 2:** 261
Demand Assigned Multiple Access (DAMA) **Volume 2:** 257
demodulation **Volume 2:** 201, 210, 212, 216, 238, 257, 275
demultiplex **Volume 2:** 221-222
Desktop Identity Certification Terminal **Volume 1:** 28
devices of military communication networks security protection system **Volume 2:** 234-235
digital auto transmissions **Volume 2:** 261
Digital Circuit Multiplication Equipment (DCME) **Volume 2:** 216, 280
Digital Code Squelch (DCS) **Volume 2:** 208
digital encryption **Volume 2:** 267, 269, 271
Digital IF Modulation **Volume 2:** 275
digital microwave communication equipment **Volume 2:** 247-248, 259
digital multiple beam forming **Volume 2:** 184
digital programmable control switch encode mode **Volume 2:** 236
digital pulse compression **Volume 2:** 185
Digital Radio Frequency Memory (DRFM) **Volume 2:** 252
digital signal transmission **Volume 1:** 50; **Volume 2:** 209
digital trunk multiplication equipment **Volume 2:** 216
Digital Video Broadcasting (DVB) link protocol **Volume 2:** 188, 197
Digital Visual Interface (DVI) **Volume 1:** 83, 85
digital-controlled processing center **Volume 1:** 55
dipole antenna **Volume 2:** 206
direct interrogation **Volume 2:** 271
direct sequence spread spectrum **Volume 2:** 267, 269, 271

direct synchronization **Volume 2:** 274
Direct-Sequence Spread Spectrum (DSSS) **Volume 2:** 247-249, 257
direction finding **Volume 2:** 180-182, 201, 210, 224, 233, 238
Director of US National Intelligence **Volume 1:** 2
disaster recovery **Volume 1:** 99
disk heartbeat **Volume 1:** 99
disperse illegal gatherings **Volume 2:** 172
distributive curve **Volume 1:** 76
diversity reception technique **Volume 2:** 282
Division Multiple Access (DMA) **Volume 2:** 208
DJG8715G Intergrated Ground-to-Air EW System **Volume 2:** 217-219
DM-15A Multiplex **Volume 2:** 230
DM-16 Delta Multiplexer **Volume 2:** 221
document management **Volume 1:** 28
domestic router **Volume 1:** 91-92
Dongfeng "Warrior" Vehicle **Volume 2:** 241
DongHu New Technology Development District of Wuhan **Volume 1:** 61
dopler radar **Volume 2:** 256
double broadband integrated digital remote camera communication equipment **Volume 1:** 78
double hot-standby operation mode **Volume 2:** 236
drug trafficking **Volume 2:** 165
DSP device **Volume 2:** 243
dual antenna **Volume 1:** 50
Dual-Tone Multi-Frequency (DTMF) **Volume 2:** 230
DVB **Volume 2:** 198
DVM4A Data and Voice Multiplexer **Volume 2:** 222
Dynamic Host Configuration Protocol (DHCP) IP address **Volume 2:** 194
dynamic portrait system **Volume 1:** 23
dynamic replacing **Volume 2:** 189
dynamic tracking performance **Volume 2:** 189
DZ9000S/DZ9001/DZ9001C/DZ9002 **Volume 2:** 223-224

E

E-government **Volume 1:** 93
E&M signaling **Volume 2:** 236
E1 circuit simulation **Volume 2:** 280
E1 fractional, E2 (8 Mbps), E3 (34 Mbps) encryption **Volume 2:** 261
E1 trunk interface **Volume 2:** 216, 243-244
E6465 **Volume 1:** 86-99
early warning detection **Volume 2:** 164, 279

earthquake **Volume 1:** 104-105; **Volume 2:** 175
East New Zone **Volume 2:** 253-254
eavesdropping **Volume 2:** 261
echo bucking function **Volume 2:** 243
echo cancelation **Volume 2:** 216
eco-system chain **Volume 1:** 93
education **Volume 1:** 28, 94
Effective Radiation Power (ERP) **Volume 2:** 182
Eight Phase Shift Keying (8PSK) **Volume 2:** 257
EIRP **Volume 1:** 43-43
elastic bearing rods **Volume 2:** 268, 272
electric grid **Volume 1:** 93
electrical vacuum equipment **Volume 1:** 74
Electro Optical (EO) **Volume 1:** 106-120; **Volume 2:** 186, 224, 237, 241, 252, 253-257
Electromagnetic (EM) **Volume 1:** 67, 79, 83, 85; **Volume 2:** 184, 223-224, 237, 257, 285
Electromagnetic Compatibility (EMC) **Volume 1:** *83, 85*
electron gun **Volume 1:** 77
electronic compass **Volume 2:** 177
Electronic Counter Measures (ECM) **Volume 1:** 65-67, 79; **Volume 2:** 180-182, 184, 252
Electronic Counter-Counter Measures (ECCM) **Volume 2:** 247-249, 259-260
Electronic Intelligence (ELINT) **Volume 1:** 1; **Volume 2:** 223-224, 233, 252
Electronic Order Of Battle (EOB) **Volume 2:** 223-225, 228, 233
Electronic Support Measures (ESM) **Volume 2:** 182, 186, 220, 233, 252
electronic tube work **Volume 1:** 74
Electronic Warfare (EW) **Volume 1:** 1, 41-43, 46, 57-60, 66; **Volume 2:** 160, 179-186, 217-220, 223, 247-248, 252, 258-259, 279
embedded monitoring terminal **Volume 2:** 259
emergency command **Volume 2:** 188
emergency key destruction **Volume 2:** 263
emulation **Volume 1:** 1; **Volume 2:** 160
encrypted chip **Volume 1:** 79
encryption **Volume 1:** 1, 50; **Volume 2:** 160, 216, 225, 228, 234, 245-248, 259, 261, 263-265, 267, 269, 270-271, 278, 282
end integrated design **Volume 1:** 96
end-to-end public security **Volume 1:** 16-17
end-to-end security solution **Volume 1:** 2
Engineering Order Wire (EOW) **Volume 1:** 50; **Volume 2:** 247-248, 259, 282, 284
enhanced casino security **Volume 1:** 24
Enterprise Linux Operating System **Volume 1:** 93

Enterprise Network Core **Volume 1:** 91-92
entertainment application **Volume 1:** 93
ERR107A Portable Radar **Volume 1:** 58
error control **Volume 2:** 264
error in data link mode **Volume 2:** 269
error of interrogation **Volume 2:** 269
ethernet **Volume 1:** 15, 83, 85-89, 91-92; **Volume 2:** 185, 197-198, 243, 280
EUROCOM **Volume 2:** 221, 230, 232, 247, 249
exchange chip **Volume 1:** 91-92
explosives **Volume 2:** 195
external clock input interface **Volume 2:** 236
extract transform and load (ETL) **Volume 1:** 26
eyeglass camera system **Volume 1:** 2

F

Face Recognition Vendor Test (FRVT) **Volume 1:** 20-21, 32-33
facial recognition **Volume 1:** 2, 15-17, 19-21, 23-24, 27-28, 32-35
Facial Recognition Prize Challenge (FRPC) under the US Office of the Director of National Intelligence and hosted by the Intelligence Advanced Research Projects (IARPA) **Volume 1:** 2, 20-21, 32-33
Factory 722 **Volume 1:** 65-68
false alarm **Volume 2:** 185
false license plate **Volume 1:** 23
far range air surveillance radar **Volume 1:** 73
Fault Tolerance (FT) CPU **Volume 1:** 93
Fax **Volume 1:** 50, 63; **Volume 2:** 209, 214, 221-222, 230-232, 236, 245, 248, 264-265, 280, 285
FH Radio Series **Volume 1:** 63
fiber optic **Volume 1:** 48, 82; **Volume 2:** 166, 168, 234, 246, 280, 285
field emergency communication **Volume 2:** 259
Field-Programmable Gate Array (FPGA) **Volume 2:** 243
fire control radar **Volume 2:** 252
fire control system (FCS) **Volume 1:** 1; **Volume 2:** 160
fire fighting **Volume 2:** 175
fire rescue **Volume 2:** 195
fire support **Volume 2:** 189
fire-control calculation **Volume 2:** 186
firewall **Volume 2:** 234-235, 262
FireWire (IEEE1394) **Volume 1:** 83, 85
fishing **Volume 2:** 165
Fixed Frequency (FF) **Volume 2:** 201, 210, 212

fixed troposcatter communication equipment **Volume 2:** 225
fixed user **Volume 2:** 250-251f
fixed wireless signal blocker model: S1-01 **Volume 2:** 208
fixed-site communications network **Volume 2:** 280
flares **Volume 2:** 252, 258
Flight Knee Tablet **Volume 1:** 78, 79, 81
FM **Volume 2:** 208, 238
FocusBrain System **Volume 1:** 2, 15-19
fog penetration **Volume 2:** 170
Foreign Exchange Office (FXO) interface **Volume 2:** 222
Foreign Exchange Subscriber (FXS) interface **Volume 2:** 222
Forward Error Correction (FEC) **Volume 2:** 247-248, 257, 259, 264
Four Integrations: 1) defense electronics system. 2) public security. 3) overseas engineering. 4) business solutions. **Volume 1:** 104-105
four-stage depressing collector **Volume 1:** 77
frame relay data **Volume 2:** 280
fratricide **Volume 2:** 266, 271, 274
Freescale Power PC P1022/T1024 **Volume 1:** 86-99
frequency agile **Volume 2:** 249
Frequency Division Multiple Access (FDMA) **Volume 2:** 208
frequency hopping **Volume 1:** 62, 64, 67; **Volume 2:** 201, 204, 209-210, 212, 229, 267, 269, 271, 274
frequency measurement **Volume 1:** 103; **Volume 2:** 180-181, 250-251
Frequency Modulated Continuous Wave (FMCW) Radar **Volume 2:** 191
Frequency-Division Duplexing (FDD) **Volume 2:** 259-260
Frequency-Division Multiple Access (FDMA) **Volume 2:** 257
Frequency-Division Multiplexing (FDM) technology **Volume 2:** 194
friendly exercise area **Volume 1:** 66
front-plug-in structure **Volume 2:** 243
frontier defense **Volume 2:** 189
frontier reconnaissance sensor terminal **Volume 1:** 66
Fudan University Shanghai Cancer Center **Volume 1:** 36-38, 40
full duplex communication mode **Volume 2:** 264
full identification mode **Volume 2:** 271, 274
full virtualization **Volume 1:** 96
full-digital IF modulation **Volume 2:** 277
fuses **Volume 2:** 208

G

G.729 CS-ACELP **Volume 2:** 216
G20 Hangzhou Summit **Volume 1:** 2, 20, 32, 34

G3 facsimile **Volume 2:** 216
G703 **Volume 2:** 243-244
G729a voice codec **Volume 2:** 222, 243
Gain Control: AGC, MGC, SMA **Volume 2:** 212
GB12401 **Volume 2:** 176
General Equipment Headquarters of the People's Liberation Army **Volume 1:** 1; **Volume 2:** 160
genetics **Volume 1:** 26
geographical tracking **Volume 1:** 112
Germany **Volume 1:** 1, 78; **Volume 2:** 160
gigabit core switch **Volume 1:** 91-92
gigabit fiber optic **Volume 2:** 234
gigabit high-speed computing network **Volume 1:** 95
GJB150.16-1986 **Volume 1:** 83, 85
GJB151A-1997 **Volume 1:** 83, 85
GJB9001 A-2001 military standard (MILSTD) **Volume 1:** 55
GL Graphics **Volume 1:** 83-84
Global Information Grid (GIG) **Volume 2:** 279
Global System for Mobile Communications (GSM) **Volume 2:** 208
GP328/GP338 radio **Volume 2:** 207
Global Positioning System (GPS) **Volume 2:** 177, 267, 269, 272, 274-275, 276-277
graphical fax service **Volume 2:** 209
graphics library **Volume 1:** 83-84
graphics processing unit (GPU) **Volume 1:** *86-99*
grid controlled phase-amplitude-matched pulsed TWT **Volume 1:** 75
ground control station **Volume 1:** 55
Ground Data Terminal (GDT) **Volume 2:** 278
ground fire-control radar **Volume 2:** 182
group-channel encryption **Volume 2:** 247-248, 282
GS-500/GS-510/GS-514 fixed troposcatter communication equipment **Volume 2:** 225-226
GS-514 Fixed Troposcatter Communication **Volume 2:** 227-228
Guangdong **Volume 2:** 175
Guangxi **Volume 2:** 175
guidance technology **Volume 1:** 1
guided-laser weapon **Volume 1:** 113
guiding aiming unit **Volume 2:** 164
Guilin Changhai Development Co. Ltd. **Volume 1:** 65
Guiyang Long Dong Bao Airport **Volume 1:** 72
gun mantlet **Volume 2:** 268
gun-pointing adjustment radar **Volume 2:** 182
gyro stabilization **Volume 1:** 107-110

H

Hainan-Guangdong-Guangxi cross-sea firefighting drill **Volume 2:** 175
half-duplex communication mode **Volume 2:** 264
Hanfan Interactive Electronic Technical Manual (IETM) **Volume 1:** 78-79
harmonics spurious **Volume 2:** 205
Hawking, Stephen **Volume 1:** 29-30
Headquarter Communication Center **Volume 2:** 250
healthcare **Volume 1:** 20, 26, 28, 36, 99
helicopter **Volume 1:** 53-55, 83-84; **Volume 2:** 170, 190-191, 215, 239
helix slow-wave **Volume 1:** 75, 77
HF Digital Transmitter **Volume 2:** 205
HF digitalized anti-jamming radio **Volume 2:** 229
HF direction finding **Volume 2:** 201
HF radio communications **Volume 2:** 264, 279
HF Single Side Band (SSB) radios communication relay **Volume 2:** 264
HF transceiver and telephone retransmission system **Volume 2:** 203
hierarchical key system **Volume 2:** 261

High Availability (HA) environment **Volume 1:** 99
high definition imaging **Volume 1:** 108
High Frequency (HF) **Volume 2:** 203, 212, 214, 229, 238, 263, 264, 279
high performance computing (HPC) **Volume 1:** 95
High Performance Computing (HPC) Cluster System **Volume 1:** 95
high speed modem **Volume 2:** 257
high speed serial bus standard **Volume 1:** 83, 85
high-speed data transmission **Volume 2:** 209
high-speed radio network **Volume 2:** 214
Hillhouse Capital Group **Volume 1:** 3, 20, 32-33
HK-BSS Border Surveillance System **Volume 2:** 165-166
HLX-240 Hot Line Tactical Switch **Volume 2:** 230-233
homeland security **Volume 1:** 21; **Volume 2:** 195
Hong Kong **Volume 1:** 3, 32-33, 36-38, 40
horizon's Voice **Volume 1:** 54
horizontal linear polarization **Volume 1:** 50
horizontal mechanical scanning **Volume 2:** 256
horizontal sidelobe **Volume 2:** 185
hospitals (Top Tier – Grade 3A) **Volume 1:** 20, 26
host monitoring system **Volume 2:** 234-235
hostile constant frequency **Volume 1:** 67
hostile tactical radio transmission **Volume 2:** 237

Hotel Self Check-In **Volume 1:** 28
hotline **Volume 2:** 230-233, 280
howitzer **Volume 2:** 241
HTTP **Volume 2:** 197
Huadong Electronics Group Co. Ltd. **Volume 1:** 78-81
Huawei **Volume 1:** 36-38, 40, 98
human rights groups **Volume 1:** 2
human-computer interaction **Volume 1:** 28; **Volume 2:** 177
human-computer interface antenna controller one-button satellite alignment **Volume 2:** 177
hybrid network **Volume 2:** 188
HZ100 shipborne ESM/ELINT system **Volume 2:** 233

I

i.Link (IEEE1394) **Volume 1:** 83, 85
IC Card **Volume 2:** 262
IC-F3GT/S radio **Volume 2:** 207
IC-F4GT/S radio **Volume 2:** 207
Identification Card Verification **Volume 1:** 28
Identification Friend or Foe (IFF) **Volume 1:** 56; **Volume 2:** 163-164, 184, 242, 266-271, 276-277
identification of in-band data **Volume 2:** 216
Identity Authentication ACS (Access Controller Access) Terminal **Volume 1:** 28, 79
Identity Authentication Security Lane **Volume 1:** 28
Identity Verification for Exams **Volume 1:** 28
IF digital sampling **Volume 2:** 277
illegal fishing **Volume 2:** 165
image processing **Volume 1:** 95
image transmission **Volume 2:** 247-248
immigration **Volume 1:** 36-38, 40
Improved Multiband Excitation (IMBE) **Volume 2:** 263, 265
in-band data **Volume 2:** 216, 280
incoming signal (Rx) frequency **Volume 2:** 198
India **Volume 1:** 2, 20, 72, 78
Indoor Base-Station (IBS) **Volume 2:** 257
industrial ecology **Volume 1:** 98
inertia navigation **Volume 2:** 189
infantry **Volume 1:** 53, 55, 58, 63-64, 67, 79; **Volume 2:** 266, 285
InfiBand **Volume 1:** 95
information and network attack **Volume 2:** 250
information retrieval **Volume 1:** 29

information security **Volume 1:** 90; **Volume 2:** 234-235
informatization **Volume 1:** 1, 64, 69-70; **Volume 2:** 160
infrared **Volume 1:** 107-109, 111-113, 116-118, 120; **Volume 2:** 180, 237, 241, 247-248, 254, 256-259
infrared circumferential scanning **Volume 2:** 256
Ingress Protection (IP) **Volume 1:** 42-43, 79, 81
initial vector **Volume 2:** 262
Institute of Electrical & Electronics Engineers (IEEE) **Volume 1:** 83-85; **Volume 2:** 214, 243-244
Integrated Node Switching Vehicle **Volume 2:** 280
integrated security management system **Volume 2:** 234-235, 279
Integrated Services Digital Network (ISDN) User Part (ISUP) **Volume 2:** 236
integrated testing instrument **Volume 2:** 246
Integration Node Switching Vehicle **Volume 2:** 280-281
Intel Xeon Processor **Volume 1:** 16, 18
Intelligence Advanced Research Projects (IARPA) **Volume 1:** 2, 20-21
intelligence diagnostic assistance system **Volume 1:** 20-21
intelligence synthesis **Volume 1:** 73
intelligent auxiliary diagnosis **Volume 1:** 25-26
intelligent diagnosis assistance products **Volume 1:** 32-34
Intelligent Medical Record Search Engine **Volume 1:** 26
IntelSat Earth Station Standards (IESS) 308/309 **Volume 2:** 176, 257
inter-link communication **Volume 2:** 259
inter-station control signal **Volume 1:** 50; **Volume 2:** 225, 282, 284
Interactive Electronic Technical Manual (IETM) **Volume 1:** 78
Intercept Civilian UAV **Volume 1:** 46
interception **Volume 1:** 67; **Volume 2:** 201, 210, 237, 257, 259
Interfaces: VGA, DVI, LVDS, PAL1553B, ARINC 429, ethernet, RS-422, IEEE1394 **Volume 1:** 83, 85
internal synchronization **Volume 2:** 243-244
International Collegiate Programming Contest (ICPC) **Volume 1:** 29
International Telecommunication Union (ITU) **Volume 2:** 222, 236, 247, 249, 259-260
Internet **Volume 1:** 63'; **Volume 2:** 188, 197, 250-251
Internet Protocol + Next-Generation Network (IP+NGN) **Volume 2:** 250-251
interrogation modes: Mode 1, 2, 3/A, C, Secure Mode **Volume 2:** 275, 277
interrogation platform **Volume 2:** 269
intra-violet radiant sensitivity **Volume 1:** 76
Intrusion Detection System (IDS) **Volume 2:** 234-235
ionosphere **Volume 2:** 225, 282
IP address **Volume 2:** 194
IP data **Volume 2:** 262, 280, 280, 284-285
IP integrated switching **Volume 2:** 280

IP packet filtering **Volume 2:** 262
IP protocol suite **Volume 2:** 214
IP switching **Volume 2:** 279
IP65 Waterproof (Ingress Protection) **Volume 1:** 79, 81
IP67 Waterproof (Ingress Protection) **Volume 1:** 46-47
IP68 Waterproof (Ingress Protection) **Volume 1:** 42-43; **Volume 2:** 202
IPSec Tunnel **Volume 2:** 262
IPv4/IPv6 **Volume 1:** 91-92
IPX-7 Waterproof **Volume 2:** 205, 209
ISX001 STM Integrated Trunk Node Switch **Volume 2:** 236
Italy **Volume 1:** 78
ITU-T (International Telecommunication Union) **Volume 2:** 222, 236, 247, 249, 259-260

J

jammer **Volume 1:** 41-47, 55-56, 66, 73; **Volume 2:** 180-182, 192-193, 219-220, 225, 229, 237, 252, 255-256, 258-259, 278, 282, 284-286
Japan **Volume 1:** 1; **Volume 2:** 160
JB Unit **Volume 2:** 203
Jiang Ning Economic District **Volume 1:** 57-60
Jiao Tong University **Volume 1:** 29
Jin Hongqiao International Center **Volume 1:** 39
Jinjiang Information Industrial Co., Ltd. **Volume 1:** 71-73
JL3D-91B Metric Wave 3D Radar **Volume 1:** 73
JN1102 UAV-Mounted Communication Countermeasures System **Volume 2:** 237
JN1118B HF/VHF/UHF Shipborne Communication Surveillance System **Volume 2:** 238
judicial **Volume 1:** 28

K

Ka-band television receive only (TVRO) **Volume 2:** 188
KDMC system **Volume 2:** 261, 264-265
Key Management Center (KMC) **Volume 2:** 262
key place guard **Volume 2:** 169
KGH (2004) No. 201 **Volume 2:** 277
kinetic strike **Volume 2:** 255
klystron **Volume 2:** 225-226
Korea **Volume 1:** 78
Ku-band **Volume 1:** 53-54; **Volume 2:** 189, 188, 191, 197, 220, 285
Kylin (Neokylin) **Volume 1:** 93-100

Kylin Cloud Desktop Management System **Volume 1:** 94-96
Kylin HPC (High Performance Computing) Cluster System **Volume 1:** 95
Kylin Information Technology Co. Ltd. **Volume 1:** 93-96
Kylin OS + FT CPU (ARM64) **Volume 1:** 93
Kylincloud Platform Management System **Volume 1:** 96
Kylinos **Volume 1:** 93-96

L

L-antenna **Volume 2:** 205-206
L-band **Volume 1:** 55-56, 77; **Volume 2:** 194, 213, 220, 257
L626 Land-based Radar EW Equipment **Volume 2:** 182
LAN **Volume 2:** 212, 236, 243-244, 246-248
Langkawi International Maritime and Aerospace Exhibition (LIMA) **Volume 2:** 182-193
LANU-M1 Vehicle Mounted UAV Jammer Vehicle **Volume 1:** 41-43
LANU-W1 Portable Wide UAV Angle **Volume 1:** 44-47
laser **Volume 1:** 1, 48, 112-120; **Volume 2:** 160, 172, 186, 241, 242, 254-256, 258
laser guided weapon **Volume 1:** 114, 117-119
laser range finder **Volume 1:** 111-120; **Volume 2:** 172
Latin America **Volume 1:** 105
law enforcement (*see* police)
Liquid Crystal Display (LCD) **Volume 1:** 78-79, 83-85
LDB-06 Battlefield Surveillance Radar **Volume 2:** 239
LDB-10 Light Reconnaissance Vehicle System **Volume 2:** 240-241
LE120 EO Payload **Volume 1:** 107
LE140 EO Payload **Volume 1:** 108
LE180 EO Payload **Volume 1:** 109
LE220 EO Payload **Volume 1:** 110
LE260 EO Payload **Volume 1:** 111
LE330D EO Payload **Volume 1:** 112
LE350 EO Payload **Volume 1:** 113
LE380 EO Payload **Volume 1:** 114
LE400 I-IV **Volume 1:** 115
LE430 EO Payload **Volume 1:** 116
LE450 EO Payload **Volume 1:** 117
LE450H EO Payload **Volume 1:** 118
LE480 I-III Model EO Turret **Volume 1:** 119
LE500 I-II EO Payload **Volume 1:** 120
LeCun Yann André **Volume 1:** 29-30
LEO Zhu **Volume 1:** 29-30
LH-2 Portable Searching And Tracking Radar **Volume 2:** 190-191
LI15AFM Military Airborne TFT LCD Monitor **Volume 1:** 83-85

Licensing as a Service (LaaS) **Volume 1:** 96
light guide control panel **Volume 1:** 78-79
LIN Cheng **Volume 1:** 29, 31
Line of Sight (LOS) **Volume 1:** 117, 119, 120; **Volume 2:** 210, 247-249, 259-260, 265, 278
Linear Frequency Modulated Continuous Wave (LFMCW) Radar **Volume 2:** 191
linear polarization **Volume 2:** 198
link mode **Volume 2:** 271
link planning software **Volume 2:** 284
link status management **Volume 2:** 259
Linux **Volume 1:** 93, 97, 98
lithium battery **Volume 2:** 197
liveness detection **Volume 1:** 27-28
logistics **Volume 2:** 279
Longwave Infrared Thermal Imager (LWIR) **Volume 1:** 103, 114-117, 119-120
Loogson Platform **Volume 1:** 98
Loong Eye **Volume 1:** 106-120
loop group in-band user signaling **Volume 2:** 221
loop relay interface **Volume 2:** 236
loop trunk **Volume 2:** 230
Low Probability of Intercept (LPI) **Volume 2:** 163, 247
low radio frequency **Volume 2:** 164
low sidelobe antenna **Volume 2:** 185
low slow small target **Volume 2:** 255-256
Low Voltage Differential Signaling (LVDS) **Volume 1:** 83, 85
Low Altitude Laser Defending System (LASS) **Volume 1:** 48
low-rate voice coding **Volume 2:** 222, 243
LSB **Volume 2:** 238
Ludi Cave Scenic Area **Volume 1:** 65
Luoyang Institute of Electro-Optical Equipment **Volume 1:** 106-120
LWE Laser Warning System Vehicle **Volume 2:** 242

M

Macau **Volume 1:** 3 32-33, 36-38, 40
machine learning **Volume 1:** 27
machine vision **Volume 1:** 23, 28
magnetic storm **Volume 2:** 225, 282
magneto telephone subscribers **Volume 2:** 230
Maipu Communication Technology Co., Ltd. **Volume 1:** 91-92
Malaysia **Volume 1:** 7, 82-10, 15-21, 23-29, 32-38, 40-50; **Volume 2:** 162-170, 179-186, 194-210, 217-219, 227-228, 240-241, 253-254

Man Machine Interface (MMI) **Volume 1:** 94; **Volume 2:** 220, 222-223, 225, 230, 243, 252, 259, 282, 286
manpack radio **Volume 2:** 194, 202, 209, 229
Manpack Transceiver Model: 2110 **Volume 2:** 202
Manportable Air Defense Missiles (MANPADs) **Volume 2:** 163
Manual Gain Control (MGC) **Volume 2:** 212
manual key distribution **Volume 2:** 261
map operation terminal **Volume 1:** 78-79
maritime **Volume 2:** 165-166, 194-195
Massachusetts Institute of Technology (MIT) **Volume 1:** 29-30
MDVR **Volume 1:** 105
mechanical phase scan system 3D radar **Volume 1:** 55-56
medicine (*see also* hospital/healthcare) **Volume 1:** 20, 25-26, 28, 95-96, 98-99
Menlo Park **Volume 1:** 39
mental test **Volume 1:** 79
mesh network **Volume 2:** 188
Message Transfer Part (MTP) **Volume 2:** 236
meteorological **Volume 1:** 72; **Volume 2:** 258
meteorological and hydrological products **Volume 1:** 72
metric wave medium/far range air surveillance radar **Volume 1:** 73
MFA01A/B Data and Voice Multiplexer **Volume 2:** 243-244
micrometer infrared band **Volume 2:** 180
microprocessor **Volume 2:** 221
Microsoft **Volume 1:** 3, 29, 31-33, 36, 38, 98, 101
Microsoft Azure Cloud Computer Platform **Volume 1:** 32-33
Microsoft Product Reaction Cards (MPRC) Platform **Volume 1:** 98
Microsoft Research Asia (MSRA) **Volume 1:** 29, 31
microwave **Volume 1:** 74-75; **Volume 2:** 163, 186, 214, 241, 246, 248, 256-257, 259-261, 279-280
microwave amplifier **Volume 1:** 74-75 ; **Volume 2:** 186
Microwave Communication Vehicle **Volume 2:** 246, 279
microwave interference **Volume 2:** 256
microwave modulation **Volume 2:** 257
microwave radio **Volume 2:** 248
Middle Wavelength Infrared (MWIR) Thermal Imager **Volume 1:** 111, 113-117. 119-120
MIL-STD-188-141B ALE **Volume 2:** 202
MIL-STD-461C **Volume 2:** 270, 272-273, 275
MIL-STD-810 **Volume 2:** 207, 270, 272-273, 275
Milestone Systems **Volume 1:** 36-38, 40
military detonation **Volume 2:** 195
military network partioning switching device **Volume 2:** 234-235

millimeter wave radar **Volume 2:** 254
Mine Clearance Vehicle **Volume 2:** 271
Ministry of Industry and Information Technology **Volume 1:** 1; **Volume 2:** 160
Ministry of Public Security (MPS) **Volume 1:** 32, 35; **Volume 2:** 175
MINPTT **Volume 2:** 236
missile **Volume 1:** 53, 66, 82; **Volume 2:** 163-164, 169, 183-184, 186, 250-252, 254, 258, 273, 286
missile communication network **Volume 2:** 250-251
missile guiders **Volume 2:** 163
missile terminal guidance **Volume 2:** 252
mixed trunk node **Volume 2:** 231
MMIC Technology **Volume 1:** 46
Mobile Banking Face Authentication **Volume 1:** 32, 35
Mobile Troposcatter Communiation Equipment **Volume 2:** 282-283
Mobile Trunk Communication System **Volume 2:** 280
MobiNet **Volume 2:** 245
Mode 1, 2, 3/A, C, Secure Mode **Volume 2:** 275, 277
modem **Volume 2:** 257, 264
modulation **Volume 2:** 201, 257
monopulse **Volume 2:** 186, 252
mortars **Volume 1:** 48 ; **Volume 2:** 241
motion blur (facial recognition) **Volume 1:** 21
Moving Target Detection (MTD) **Volume 2:** 184, 186
Moving Target Indication (MTI) processing **Volume 2:** 184
moving-to-moving **Volume 2:** 267, 269, 271
moving-to-static **Volume 2:** 267, 269, 271
MS-700 Battlefield Switch Series **Volume 2:** 245-246
multi-band test **Volume 1:** 103
multi-clock synchronization mode **Volume 2:** 221
multi-diversity receiving technology **Volume 2:** 228
Multi-Flexing Time Division Multiple Access (MF-TDMA) **Volume 2:** 188, 197
multi-information-rate **Volume 2:** 259
multi-processing distribution communication **Volume 2:** 243
Multi-Purpose Vehicle (MPV) **Volume 2:** 177
Multiple Channels Per Carrier (MCPC) General Modem **Volume 2:** 257
Multiple Spectrum Photoelectric Reconnaissance System (TH-S711/PS) **Volume 2:** 169
Multiprotocol Label Switching (*MPLS*) **Volume 1:** 91-92; **Volume 2:** 245
multiwave countermeasure **Volume 2:** 254
MW-800 **Volume 2:** 247, 249
MW-1500 **Volume 2:** 247, 249
MW-4800 **Volume 2:** 247, 249

MW-Series Tactical Microwave Radio **Volume 2:** 247-249
MyPower NSS6600 **Volume 1:** 91-92

N

N Connector **Volume 2:** 194
N-F coaxial output **Volume 1:** 75
n:N Comparison **Volume 1:** 23
Nanjing Changjiang Electronics Group Co. Ltd. (NJCJEC) **Volume 1:** 53-54, 55-56
Nanjing Corad Electronic Equipment Co., Ltd. **Volume 1:** 57-60
Nanjing Economic Development Zone **Volume 1:** 55-56
Nanjing Huadong Electronics Group Co. Ltd. **Volume 1:** 78-81
Nanjing Panda Handa Technology Co. Ltd. **Volume 1:** 53-54
Nanjing Sanle Electronics Corp **Volume 1:** 74-77
narrow-beam directional transmission **Volume 2:** 228, 225, 282
narrowband **Volume 2:** 280
National Communication System **Volume 2:** 250
National Defense Trunk Link **Volume 2:** 284
National Defense University **Volume 2:** 175
National Development and Reform Industrialization Special Support **Volume 1:** 94
National Institute of Standards and Technology (NIST) under the US Department of
    Commerce **Volume 1:** 2, 20-21
National Key New Products Award **Volume 1:** 97
National Military Communication Solution **Volume 2:** 250-251
National Research Laboratory **Volume 1:** 74
National Security Asia (NATSEC Asia Exhibition/Defence Services Asia/Malaysia)
Volume 1: 3
Natural Language Processing (NLP) **Volume 1:** 26
Naval Communication Network **Volume 2:** 250-251, 279
navigation **Volume 1:** 56; **Volume 2:** 180-181, 189, 219, 256, 258
navy **Volume 1:** 53, 55, 68, 79, 82, 118-119; **Volume 2:** 164, 182, 184, 190, 225, 233,
    252, 258, 264, 276-277, 282
NeoKylin HA (high availability) Cluster **Volume 1:** 99
NeoKylin Operating System **Volume 1:** 97-98
NeoShine **Volume 1:** 101
network management **Volume 2:** 188, 216, 236, 246, 259, 279, 280-282
Network Management Center Vehicle **Volume 2:** 280-281
Network Management System (NMS) **Volume 2:** 236
Network Management Vehicle **Volume 2:** 246
network protocols **Volume 1:** 91-92
network synchronization **Volume 2:** 232
New Poly Plaza **Volume 1:** 16, 18, 50

New Wheeled Armored Tactical Vehicle (YJ2080C) **Volume 2:** 171-172
New York University (NYU) **Volume 1:** 29-30
Next Generation Network (NGN) **Volume 2:** 279
NGT SR radio **Volume 2:** 207
NiMH Battery **Volume 2:** 202
NLP Technology (natural language processing) **Volume 1:** 26
NMS **Volume 2:** 246, 250-251
non-block fully-digital time division switching **Volume 2:** 230
non-block information switch **Volume 2:** 236
non-blocking PCM (phase change material) switching matrix **Volume 2:** 230
non-communication emitters **Volume 2:** 233
Non-Frontal Head Pose (facial recognition) **Volume 1:** 21
non-inter connect technique **Volume 2:** 243
Non-Neutral Facial Expressions (facial recognition) **Volume 1:** 21
non-offensive platform **Volume 2:** 271
non-repudiation **Volume 2:** 234
NORINCO (see China North Industries Group Corp.)
NRJ5A Shipborne ESM/ECM System **Volume 2:** 252
NSR2900 self-controllable domestic router **Volume 1:** 91-92
NSS6600 **Volume 1:** 91-92
nuclear electronic equipment **Volume 1:** 1; **Volume 2:** 160
nuclear explosion **Volume 2:** 225, 282
nuclear power plant **Volume 1:** 68

O

O-Shield Comprehensive Electro-Optical Defense System **Volume 2:** 253-256
Oceania **Volume 1:** 3, 32-33
octoploid diversity **Volume 2:** 225-226
octuple diversity receiver techniques **Volume 1:** 50
Offline Key Management **Volume 2:** 262
offshore drilling **Volume 2:** 228
oil pipeline **Volume 1:** 68
Olympic security **Volume 2:** 175
omindirectional antenna **Volume 2:** 277
omnidirectional data **Volume 2:** 271
omnidirectional radar **Volume 1:** 60
omnidirectional transmission **Volume 2:** 269, 271
On-The-Move Satcom Vehicle **Volume 2:** 188-189
one-button-switch operational mode **Volume 2:** 285
one-hop **Volume 2:** 284
Online Key Management **Volume 2:** 262, 265

open architecture **Volume 1:** 83-84; **Volume 2:** 250
Open GL13/Open GL2.0 **Volume 1:** 83-84, 86-88
Open Media Application Platform (OMAP) **Volume 2:** 236
Open Platform IP Video Management Software **Volume 1:** 36-38, 40
Open Systems Interconnection (OSI) protocol stack **Volume 2:** 261
operation and command software package **Volume 2:** 164
opposite-end user **Volume 2:** 221
optical equipment **Volume 2:** 160
optical protection partial system **Volume 2:** 255
optical systems **Volume 1:** 1, 48
optoelectronic camera **Volume 2:** 166
Oracle11g.12c **Volume 1:** 98
over-the-horizon voice **Volume 1:** 55
over-air-rekeying **Volume 2:** 264-265

P

P1022 **Volume 1:** 86-99
Paired Carrier Multiple Access (PCMA) **Volume 2:** 257
pairing on the move **Volume 2:** 189
Pakistan-China Combined Operations Exercises **Volume 2:** 175
PAL1553B **Volume 1:** 83, 85
Panda Handa Technology Co. Ltd. **Volume 1:** 53-54
panoramic spectrum **Volume 2:** 201
para virtualization **Volume 1:** 96
parabolic **Volume 2:** 176, 198, 225-226, 228, 247, 249, 259-260, 282-283
parallel compilation **Volume 1:** 95
parallel computing **Volume 1:** 95
parallel digital filtering **Volume 2:** 257
parameter information **Volume 2:** 177
partial discharge (PD) signal **Volume 2:** 233
passive electronic countermeasures **Volume 2:** 180-181
passive jammers **Volume 2:** 180
passive target positioning **Volume 1:** 112
PCI eXtensions for Instrumentation (PXI) **Volume 1:** 103
pelagic region front **Volume 1:** 66
People's Armed Police (PAP) **Volume 1:** 53
People's Liberation Army (PLA) **Volume 1:** 1, 82; **Volume 2:** 160
People's Liberation Army (PLA) Rocket Force (PLARF) **Volume 1:** 82
person of interest (*also see* blacklist) **Volume 1:** 2
Personal Handy-Phone System (PHS) **Volume 2:** 208
Phase Change Material (PCM) equipment **Volume 2:** 230, 284

Phase-Amplitude-Matched Pulsed TWT BM-2021F **Volume 1:** 75
Phase-Locked Loop (PLL) **Volume 2:** 243-244
phase-scan **Volume 2:** 184
photo-electric **Volume 1:** 76; **Volume 2:** 168-169
photocathode **Volume 1:** 76
photoelectric joint reconnaissance system **Volume 2:** 168-169
phytium CPU **Volume 1:** 93, 96
pilot frequency squelch **Volume 2:** 205
pirates **Volume 2:** 190
pitch electric scanning **Volume 2:** 256
planar antenna **Volume 2:** 176, 184
platform application technology research **Volume 1:** 90
platform timing **Volume 2:** 269
Platform-as-a-Service (PaaS) **Volume 1:** 96
plug-out on line **Volume 2:** 216
point-to-multipoint application **Volume 2:** 216
Point-To-Point (P2P) subscriber switch **Volume 2:** 221-222, 243, 284
point-to-point BLOS (Beyond Line of Sight) **Volume 1:** 50
Poisson Distribution System (probability distribution) **Volume 2:** 216
polar light **Volume 2:** 225
police **Volume 1:** 2, 20, 23, 28, 36-38, 40, 98, 105; **Volume 2:** 175, 190, 195
Poly Defense **Volume 1:** 49-50; **Volume 2:** 217-219
Poly Technologies **Volume 1:** 2, 11-20, 41-48, 50
portable battlefield surveillance radar **Volume 1:** 68
portable communication jammer **Volume 1:** 66-67
portable ethernet switch **Volume 1:** 89
Portable Facial Recognition System **Volume 1:** 15
Portable Radio Integrated Tester **Volume 1:** 103
Portable Settlement Terminal **Volume 1:** 28
Power over Ethernet (PoE) **Volume 1:** 15, 18
Power Supply Vehicle **Volume 1:** 69-70
PR-11A First Precipitation Monitoring Radar **Volume 1:** 72
PRC-2000D VHF Handheld Frequency Fixed Radio **Volume 1:** 64
PRC-2000G VHF Manpack Narrow-band and High Speed Frequency Hopping Radio **Volume 1:** 63
PRC-2000H VHF Handheld Frequency Hopping Radio **Volume 1:** 64
PRC-2000S VHF Handheld Frequency Fixed Radio **Volume 1:** 64
precise servo control **Volume 2:** 189
precise time synchronization **Volume 2:** 267, 269
prison security **Volume 1:** 28, 46
Private Branch Exchange (PBX) **Volume 2:** 243
Program 863 (State High-Tech Development Plan) **Volume 1:** 94

proprietary binocular liveness detection **Volume 1**: 27
proprietary key exchange protocol **Volume 2**: 262
proprietary path tracking technology **Volume 1**: 24
pseudo-color output **Volume 1**: 109
psychological intelligence assessment and training **Volume 1**: 78
public safety **Volume 1**: 22-23
Public Switch Telephone Network (PSTN) **Volume 2**: 203, 236, 264-265
Pukou **Volume 1**: 74
pulse amplitude **Volume 1**: 76
pulse compression technique **Volume 1**: 55-56
pulse dopler radar **Volume 2**: 256
Pulse Repetition Frequency (PRF) Radar Staggering/Jittering **Volume 2**: 233
Pulse-Code Modulation (PCM) **Volume 2**: 243-244
PXI Bus Architecture (PCI eXtensions for Instrumentation) **Volume 1**: 103

Q

Qinghua International Technology Community Center **Volume 1**: 39
Quadrature Amplitude Modulation (QAM) **Volume 2**: 259-260
Quadrature Phase Shift Keying (QPSK) **Volume 2**: 257, 259-260
Quality of Service (Qos) Guarantee **Volume 2**: 197, 214
quasi-synchronization technology **Volume 2**: 230

R

radar **Volume 1**: 1, 48, 55-56, 58, 60, 62, 65-66, 68, 71-73, 79; **Volume 2**: 160, 163, 166, 168, 170, 182-184, 186, 190-191, 219, 224, 233, 241, 248, 252, 254, 256, 258-259, 267, 269, 271, 274, 285
radar-guided weapon **Volume 2**: 252
radio **Volume 1**: 62-64, 67, 103; **Volume 2**: 193, 195, 201, 203-205, 207-209, 214, 216, 222, 225, 229-230, 247-248, 256, 259-260, 263-265, 280, 282,
radio channel error correction **Volume 2**: 280
radio controlled fuses **Volume 2**: 208
radio digital communication system **Volume 2**: 282
radio relay **Volume 2**: 247-248, 259
Range Gate Pull-Off (RGPO) **Volume 2**: 182
range-finder **Volume 1**: 112; **Volume 2**: 172, 241
Real Time Streaming Protocol (RTSP) video/audio **Volume 2**: 194
real-time alarm **Volume 1**: 12, 15-17, 60
real-time channel BER test **Volume 2**: 225, 282
real-time monitoring **Volume 2**: 213
real-time telecontrol **Volume 2**: 213

real-time telemetry **Volume 2:** 213
real-time transmission **Volume 2:** 278
REC-1/1S Shipborne Search Radar **Volume 1:** 56
reconnaissance **Volume 1:** 58, 66, 119-120; **Volume 2:** 164, 167-169, 174-175, 182, 201, 213, 219-220, 239, 258, 278-279
reconnoitered intelligence **Volume 2:** 239
record of reconnaissance information **Volume 2:** 213
regional communication node **Volume 2:** 250
REL-2A Coast Sea and Low Air Surveillance Radar **Volume 1:** 56
REL-4 Long Range 3D Radar **Volume 1:** 55-56
relay transmission **Volume 2:** 280
remote access communication **Volume 2:** 285
remote control **Volume 1:** 55-56; **Volume 2:** 256
remote identity verification **Volume 1:** 27
remote sensing **Volume 1:** 1
remote sensor **Volume 2:** 160, 174-175
resource pooling management **Volume 1:** 96
resource survey **Volume 1:** 95
RF agile **Volume 2:** 252
RF coverture **Volume 2:** 186
RF input **Volume 1:** 75
RF interfaces **Volume 2:** 198
RF loops **Volume 2:** 247-248, 259
RF Signal Generation **Volume 1:** 103
rocket system **Volume 1:** 48
route reconstruction capabilities **Volume 2:** 280
router **Volume 1:** 91-92
RS-232 **Volume 2:** 270, 272-273, 243-244
RS-422 (TIA/EIA-422) **Volume 1:** 83, 85 ; **Volume 2:** 185, 243-244, 257, 270, 272-273
RS-485 (TIA-*485*/EIA-*485)* Asynchronous Semi-Duplex Bus Transmission Mode **Volume 1:** 64
RTU-292 Radio/Telephone Interface **Volume 2:** 203
Russia **Volume 1:** 2, 20
Rx (Incoming Signal) frequency **Volume 2:** 198, 213
Rx Frequency **Volume 1:** 50

S

S-band **Volume 2:** 184, 220
S1-01 Fixed Wireless Signal Blocker **Volume 2:** 208
S2/DVB-RCS+ **Volume 2:** 197

safeguard stability **Volume 2:** 172
salt-fog-proof **Volume 2:** 277
Sanle Electronics Corp. **Volume 1:** 74-77
Satcom-On-The-Move **Volume 2:** 189
satellite **Volume 1:** 54-55, 77, 95; **Volume 2:** 174-177, 188-189, 196-197, 215, 221-222, 243, 246-248, 256-257, 259, 261, 267, 269, 272, 274, 279, 285-286
    satellite communication modem **Volume 2:** 257
    Satellite Communication Vehicle **Volume 2:** 246
    Satellite Communications (SATCOM) **Volume 2:** 188-189, 230, 246, 257, 279, 285
    satellite remote sensing **Volume 1:** 95
SatTour 300A **Volume 2:** 196-197
SatTour 600Z **Volume 2:** 197
Science and Technology Progress Award **Volume 1:** 32, 35
SCM2006 burst modem **Volume 2:** 257
SCMM 500 IP Modem **Volume 2:** 257
SCPC terminal **Volume 2:** 197
Seawatch Passive Electronic Optic Jamming System **Volume 2:** 258
SEC-30 Bulk Encryptor **Volume 2:** 261
SEC-34 IP Encryptor **Volume 2:** 262
SEC-35 AM Manpack Radio Encyptor **Volume 2:** 263-264
SEC-83 VHF Shipborne Radio Encryptor **Volume 2:** 265
Second Artillery Corps **Volume 1:** 53, 82 (*see also* People's Liberation Army Rocket Force/PLARF)
second level switch **Volume 1:** 99
secure gateway **Volume 2:** 234-235
security authentication **Volume 1:** 79
security management center **Volume 2:** 246
seismic activity **Volume 1:** 76
self-check function **Volume 2:** 221, 243
self-controllable architecture **Volume 1:** 90
self-defining data **Volume 2:** 267, 271
self-exclusion scheme **Volume 1:** 24
self-loop function **Volume 2:** 221, 243
self-pairing **Volume 2:** 189
semitransparent bialkali photocathode **Volume 1:** 76
sensitive region surveillance **Volume 2:** 190
Sensitivity Time Control (STC) **Volume 2:** 275
Sensor Control Module (SCM) 8000 **Volume 2:** 257
Sequoia Capital (US) **Volume 1:** 3, 20, 32, 35
Series A Funding **Volume 1:** 20, 35
Series B Funding **Volume 1:** 20, 32, 34
Series C Funding **Volume 1:** 3 20, 32-33

Series Tactical Microwave Radio (SDR) **Volume 2:** 259-260
servo control subsystem **Volume 2:** 191
Set-Based Discrimination Subspaces Learning (SDSL) **Volume 2:** 284
Shanghai Cancer Center **Volume 1:** 36-38, 40
Shanghai Jiao Tong University **Volume 1:** 29
Shanghai Jin Hongqiao International Center **Volume 1:** 39
Shanghai Pudong Development Bank **Volume 1:** 32, 35
Shanghai Yitu Network Technology Co Ltd (see Yitu Technology)
shaped beam transmission **Volume 2:** 184
shell-burst spot **Volume 2:** 241
Shenofeifan Electronic System Technology Development Co. Ltd. **Volume 1:** 69-70
Shenzhou **Volume 1:** 82
ship cannon system **Volume 1:** 56
shipborne radar **Volume 2:** 182
shipborne transponder **Volume 2:** 276-277
shipping verification **Volume 1:** 28
ships **Volume 1:** 55-56, 68-70, 118; **Volume 2:** 164-166, 184, 190, 195, 205, 209, 229, 233, 239, 241, 252, 258, 264-265, 276-277
short identification mode **Volume 2:** 271, 274
shortwave **Volume 1:** 53, 69-70, 79, 103
Sichuan Province **Volume 2:** 175
side-band suppression **Volume 2:** 205
sideslope angles **Volume 2:** 269
signal demodulation **Volume 1:** 103
signal parameter **Volume 2:** 201, 224
signal processing subsystem **Volume 2:** 163, 191
signal tracking **Volume 2:** 198
Signal-To-Noise and Distortion (SINAD) **Volume 2:** 205, 207
Signaling Connection Control Part (SCCP) **Volume 2:** 236
Signaling System: MTP2/P3, TUP, ISUP, SCCP, TCAP **Volume 2:** 236
signals anti-interception capabilities **Volume 2:** 257
Silent Hunter Vehicle counter-UAS laser weapon (Poly Technologies) **Volume 1:** 48
silent voice processing **Volume 2:** 216
Simple Network Management Protocol (SNMP) **Volume 2:** 216
Singapore **Volume 1:** 3, 32-33, 36-39, 40; **Volume 2:** 173-178
Singapore Airshow **Volume 2:** 173-178
Single Carrier Modulation (SCM) **Volume 2:** 257
Single-Carrier Phase Coding (SCPC) **Volume 2:** 188
single-hop distance **Volume 2:** 225, 282
single-offset feed parabolic antenna **Volume 2:** 225-226
single-point malfunction **Volume 1:** 99
Sm-Co Electro-Magnetic Focusing System **Volume 1:** 75, 77

SMA-F /K **Volume 1:** 75
Smart ACS (Access Controller Access) **Volume 1:** 28
smart bracelet **Volume 1:** 78
Smart Casino **Volume 1:** 24
Smart City **Volume 1:** 3, 22, 32-33
Smart Finance **Volume 1:** 27
Smart Glasses **Volume 1:** 2
Smart Hardware **Volume 1:** 28
Smart Healthcare Solutions **Volume 1:** 25-26
Smart Security Solutions **Volume 1:** 23
SmartEye **Volume 2:** 162-164
SmartGuard TH-BS08 Ground Surveillance and Command System **Volume 2:** 167-170
SmartHunter Plus: A Portable Air Defense Missile Operation and Command System (TH-S311A) **Volume 2:** 162-164
smoke grenades **Volume 2:** 258
smoke screen interference **Volume 2:** 242, 258
smuggling **Volume 2:** 190, 239
Software as a Service (SaaS) **Volume 1:** 96
software radio architecture **Volume 2:** 193
Sogou **Volume 1:** 94
solar loading **Volume 2:** 198
Soldier's Mental Status Intelligent Evaluating and Training **Volume 1:** 79
Solid State Power Amplifier (SSPA) **Volume 2:** 225-226, 228
solid-state amplification **Volume 2:** 184
source code **Volume 1:** 98
South Africa **Volume 1:** 1, 2, 20; **Volume 2:** 160
South America **Volume 1:** 104-105
South Korea **Volume 1:** 1, 78; **Volume 2:** 160
Southeast Asia **Volume 1:** 32-33, 36-38, 40
Southwest Institute of Electronics Technology (SWIFT) **Volume 2:** 266-277
space vehicle **Volume 1:** 76, 82-84
spatial energy synthesis **Volume 2:** 184
special communication system **Volume 2:** 189
Special Forces **Volume 2:** 225, 282
special military networks **Volume 2:** 280
special movable applications **Volume 2:** 282
spectral response characteristics **Volume 1:** 76
speech recognition **Volume 1:** 28
Sports Utility Vehicle (SUV) **Volume 2:** 177
SR2410C 3D Multi-Functional Radar **Volume 2:** 184-185
SR64 Search Radar Vehicle **Volume 2:** 183

ST-16A Vehicular Combined Interrogator Transponder **Volume 2:** 266-270
ST-16B Vehicular Transponder **Volume 2:** 271-272
ST-16H Portable Interrogator **Volume 2:** 273-274
ST-36A Interrogator **Volume 2:** 275-277
ST-410 series Anti-jamming TT&C and Communication System **Volume 2:** 278
Standard 6U CPC1 Card **Volume 1:** 86-99
standard data interface **Volume 2:** 185
star network **Volume 2:** 188
Star Online The (Malaysia) **Volume 1:** 2
State Grid Electricity **Volume 1:** 93
State High-Tech Development Plan **Volume 1:** 94
State Security Department **Volume 1:** 53
State Software Industrial Layout **Volume 1:** 97
State-run Factory 722 **Volume 1:** 65-68
static IP address **Volume 2:** 194
static portrait system **Volume 1:** 23
static-to-moving **Volume 2:** 267, 269, 271
static-to-static **Volume 2:** 267, 269, 271
stationary satellite communication vehicle **Volume 2:** 177
statistical modeling **Volume 1:** 29-30
statistical multiplexing techniques **Volume 2:** 216
stealth **Volume 1:** 73
storms **Volume 2:** 282
Strategic Communication System (SCS) **Volume 2:** 279
strategic mobile communication system **Volume 2:** 280-281
submarines **Volume 1:** 69-70, 119; **Volume 2:** 276-277
Subminiature version-A (SMA) **Volume 2:** 212
subscriber switch **Volume 2:** 231
sunspot **Volume 2:** 225
Sunway Platform **Volume 1:** 98
supercomputer **Volume 1:** 95
Support TCP/HTTP **Volume 2:** 197
surveillance **Volume 1:** 1-2 15-18, 55-56, 58, 68, 73; **Volume 2:** 165-170, 184-185, 190, 212, 237-238, 252
switch **Volume 2:** 203, 230-236, 243, 245-246, 280, 285
Switzerland **Volume 1:** 1; **Volume 2:** 160
symmetrical audio frequency (AF) **Volume 2:** 212
synchronization error **Volume 2:** 269, 272, 274
Synchronous Transfer Module (STM) **Volume 2:** 236, 246, 280
Synthetic Aperture Radar (SAR) **Volume 1:** 66
System Center Configuration Manager (SCCM) 2000 **Volume 2:** 257

T

T1024 **Volume 1:** 86-99
tablet computer **Volume 1:** 78
tactical integrated access node **Volume 2:** 250-251, 279
tactical Internet **Volume 1:** 63; **Volume 2:** 215
tail-connection communication **Volume 2:** 248, 259
tanks **Volume 1:** 64; **Volume 2:** 239, 241, 266, 268
Tanner-Whitehouse (TW3) **Volume 1:** 25
Taohua River **Volume 1:** 65
target extraction **Volume 2:** 185
taxation **Volume 1:** 28
TC-527A Photo-Electric Detector **Volume 1:** 76
Telecommunications Industry Association (TIA) **Volume 1:** 64, 83, 85
telemetry **Volume 2:** 212, 213, 257, 278
Telemetry, Tracking and Command (TT&C) **Volume 2:** 212, 278
telephone switch equipment **Volume 2:** 236
Telephone User Part (TUP) **Volume 2:** 236
Television Receive Only (TVRO) **Volume 2:** 188
Tellis-Coded Modulation (TCM) coding **Volume 2:** 264
Tencent **Volume 1:** 36-38, 40
terminal display subsystem **Volume 2:** 191
terrain following radar **Volume 1:** 219
terrorism **Volume 1:** 1, 48; **Volume 2:** 169, 172, 190, 239
TFT LCD Monitor **Volume 1:** 83-85
TH-BS08 **Volume 2:** 167-170
TH-S311A **Volume 2:** 162-164
TH-S711/PS **Volume 2:** 169
thermal image **Volume 1:** 109; **Volume 2:** 166, 241
third order intermodulation **Volume 2:** 205
Three Operating Modes: short identification, full identification, data exchange, omnidirectional link mode **Volume 2:** 267
three ring antenna **Volume 2:** 206
TIA-*485*/EIA-*485* (RS-485) **Volume 1:** 64
TIA/EIA-422 **Volume 1:** 83, 85
Tiangong Spacelab **Volume 1:** 82
Tianjin Explosion Incident **Volume 2:** 175
Tianjin Kylin Information Technology Co. Ltd. **Volume 1:** 93-96
Time Division Multiple (TDM) **Volume 2:** 188, 193, 232, 245
time hopping **Volume 2:** 264-265, 267, 269, 271, 274
time stamp **Volume 2:** 264-265
time synchronization **Volume 2:** 271

Time-Division Multiple Access (TDMA) **Volume 2:** 197
Time-Division Multiplexing (TDM) **Volume 2:** 230
Times Tech **Volume 2:** 173-178
TNC Coaxial Output **Volume 1:** 77
Top Tier – Grade 3A Hospital **Volume 1:** 20, 26, 32, 34
TR47C Tracking Radar Vehicle Borne **Volume 2:** 186
track guest **Volume 1:** 24
tracking and locating data transmission **Volume 2:** 278
tracking high speed/maneuverability targets **Volume 1:** 73
traffic **Volume 1:** 22, 28
train stations **Volume 1:** 23
training algorithm **Volume 2:** 264-265
Transaction Capabilities Application Part (TCAP) **Volume 2:** 236
transceiver **Volume 2:** 191, 207
Transmission Control Protocol (TCP) **Volume 2:** 197
Transmitted Signal (Tx) frequency **Volume 2:** 198
transparent data **Volume 2:** 280
transparent transmission **Volume 2:** 257
transparent voice **Volume 2:** 216
Transport Vehicle **Volume 2:** 271
Transportable SAR Radar Jammer **Volume 1:** 66
Traveling Wave Tube (TWT) amplifier solid state transmitter **Volume 2:** 186
Treasury Management **Volume 1:** 28
Tri-Service Defense and Security Exhibition (Thailand) **Volume 2:** 212-216, 221-226, 229-231, 233-236, 238-239, 242-252, 255-285
tripod **Volume 1:** 58, 68; **Volume 2:** 210
troposcatter **Volume 1:** 49-50; **Volume 2:** 214, 216, 221-222, 225-228, 230, 243, 246-248, 259, 279-280, 284-285
truck vehicle **Volume 2:** 170
trunk **Volume 2:** 216, 225, 230-231, 236, 243-244, 247-248, 259, 261, 280, 282
trunk communication node networking application **Volume 2:** 230, 236, 282
trunk communication system **Volume 2:** 236
trunk LOS transmission **Volume 2:** 259
trunk side interface: E1 **Volume 2:** 216
TS-501/TS-504 Mobile Troposcatter etc. **Volume 2:** 282-283
TS-504 Troposcatter Communication Vehicular Station **Volume 1:** 49-50
TS-534 Troposcatter Communication Vehicle **Volume 2:** 284
TS-802 TS/SATCOM Dual-Mode Communication Vehicular Station **Volume 2:** 285
Turbo Product Code (TPC) **Volume 2:** 257
Tv **Volume 1:** 111, 113-114, 116-117, 119-120; **Volume 2:** 254, 258
TW3 **Volume 1:** 25
twinkle jamming **Volume 2:** 182

two-way compatible mainstream office software **Volume 1:** 101
two-wire analog data **Volume 2:** 243
TWT BM-2021F **Volume 1:** 75
Tx (transmitted signal) frequency **Volume 1:** 50; **Volume 2:** 198, 202, 213, 228

U

Ukraine **Volume 1:** 1; **Volume 2:** 160
Ultra High Frequency (UHF) **Volume 1:** 62; **Volume 2:** 194, 207-208, 210, 212, 215, 229, 237-238, 278
ultra-shortwave **Volume 1:** 53, 103
ultra-strong connectivity **Volume 2:** 194
ultrasonic **Volume 1:** 25
Uneven Illumination (facial recognition) **Volume 1:** 21
Uninterruptable Power System (UPS) **Volume 2:** 188, 284
United Kingdom **Volume 1:** 1; **Volume 2:** 160
United Technologies **Volume 1:** 36-38, 40
University of California – Los Angeles (UCLA) **Volume 1:** 29-30
University of Science and Technology of China **Volume 2:** 175
Unmanned Aerial Vehicle (UAV) **Volume 1:** 2, 41-46, 48; **Volume 2:** 174-175, 191-192, 237, 278-279
Unmanned Aerial Vehicle (UAV) Technical School within the Dept. Civil Affairs of Sichuan Province **Volume 2:** 175
urban planning **Volume 1:** 22
US Department of Commerce **Volume 1:** 2, 20
US Department of Homeland Security **Volume 1:** 21
US Office of the Director of National Intelligence **Volume 1:** 2, 20-21
USB interface **Volume 2:** 197

V

V35 **Volume 2:** 243-244
vacuum electronic industry **Volume 1:** 74
vector fonts of Chinese characters **Volume 1:** 83-84
Vehicle Mobile Very Low Frequency (VLF) transmission system **Volume 1:** 69-70
vehicle recognition system **Volume 1:** 23
vehicle refitting system **Volume 2:** 174
venetian bline type multiplier system **Volume 1:** 76
Vertical Identity Authentication Terminal **Volume 1:** 28
Very High Frequency (VHF) **Volume 1:** 62; **Volume 2:** 203-205, 207-208, 210, 212, 214-215, 229, 237-238, 265
VHF Direction Finding **Volume 2:** 210

VHF FH hand-held radio **Volume 2:** 205
VHF frequency-hopping radio **Volume 2:** 204
VHF-UHF Handheld Transceiver **Volume 2:** 207
vibration **Volume 1:** 83, 85; **Volume 2:** 168, 170
vibration and acoustic monitoring system **Volume 2:** 170
VIC-2001H (PVIC-20001H) **Volume 1:** 64
Video Graphics Array (VGA) **Volume 1:** 83, 85
video interface **Volume 2:** 243-244
Video RAM (VRAM) **Volume 1:** 86-88
Video Teller Machine (VTM) **Volume 1:** 32, 35
Video-LAN Compatibility (VLC) **Volume 2:** 194
Vietnam **Volume 1:** 78
VIP Convoy Ammunition Vehicle **Volume 1:** 41-43
VIP Guest Recognition **Volume 1:** 24, 28
VIP Security **Volume 2:** 208
Virtual Private Network (VPN) **Volume 2:** 262
virtualized containers **Volume 1:** 96
VLF Power Amplifier Vehicle **Volume 1:** 69-70
Vmware **Volume 1:** 98
voice coding technique **Volume 2:** 221, 243, 263
voice input modulation response **Volume 2:** 204
voice suppression **Volume 2:** 243
voltage measurement **Volume 1:** 103
Voltage Standing Wave Ratio (VSWR) **Volume 2:** 205
Volution via Modem **Volume 2:** 264
VRC-2000 VHF Vehicular Frequency Hopping Radar **Volume 1:** 62
VRC-2000A VHF Vehicular Low Cost Fixed – Fequency Radio **Volume 1:** 62
VRC-2000G VHF Vehicular Narrow Band and Frequency Hopping Radar **Volume 1:** 62
VRC-2000L VHF Vehicular Low Cost Frequency Hopping Radio **Volume 1:** 62
VRC-2001 HF Vehicular Frequency Hopping Radar **Volume 1:** 62
VRC-3030 UHF Vehicular High Speed Radar **Volume 1:** 62
VST ECS Phil. **Volume 1:** 36-38, 404
vxWorks6.9 **Volume 1:** 86-99

W

Wanda Group **Volume 1:** 36-38, 40
war-time algorithm **Volume 2:** 264-265
waterproof **Volume 1:** 79, 81; **Volume 2:** 205, 209, 259, 277(*see* Ingress Protection or IP65/IP67/IP68)
wavelength peak **Volume 1:** 76

weather **Volume 1:** 72, 95, 111, 114-115, 120; **Volume 2:** 164-165, 168, 190, 198, 202, 239, 255, 278-279, 282, 285
Weilaikeji-cheng **Volume 1:** 69-70
Wenchuan earthquake **Volume 2:** 175
Western High-tech Zone **Volume 2:** 178
whip antenna **Volume 2:** 205
Whitehouse **Volume 1:** 25
wideband **Volume 2:** 186, 280
WiFi **Volume 1:** 2, 15, 18; **Volume 2:** 197-198
Windows **Volume 2:** 236
wireless ad-hoc network communication terminal **Volume 2:** 194-195
wireless image transmission system **Volume 2:** 174-175
wireless transmission among nodes **Volume 2:** 194
wiretapping **Volume 2:** 208
WPS **Volume 1:** 94
Wuhan **Volume 1:** 61

X

X-band **Volume 1:** 72; **Volume 2:** 220, 239, 256
X-band Portable Dual Polarization Weather Radar System **Volume 1:** 72
X86 Platform **Volume 1:** 98
Xeon Processor **Volume 1:** 2
Xiangji-Haidun Technology Co. Ltd. **Volume 1:** 82-89
Xinlong Electronics New Technology Co. Ltd. **Volume 1:** 1; **Volume 2:** 160
XS-3 Remote Broadband Tactical Communication System based on Ad Hoc Network **Volume 2:** 192-193

Y

Ya'an Lushan earthquake **Volume 2:** 175
Yanjing Auto **Volume 2:** 171-172
Yitu Technology **Volume 1:** 2-40
Yitu-Beijing **Volume 1:** 39
Yitu-Hangzhou **Volume 1:** 39
Yitu-Shanghai **Volume 1:** 39
Yitu-Singapore **Volume 1:** 39
Yuille, Alan **Volume 1:** 29-30
Yunfeng Capital **Volume 1:** 3, 20, 32, 34
YWT-1 Command and Communication System **Volume 2:** 286

Z

Z interface **Volume 2:** 243-244
zero latency **Volume 2:** 261
Zhaoxin Platform **Volume 1:** 98
Zhejiang Fortune Center **Volume 1:** 39
Zhenfund **Volume 1:** 3, 20, 32, 35
Zhu LEO **Volume 1:** 29-30
Zhuhai Airshow **Volume 1:** 53-60, 65, 69-89, 91-97, 102-105; **Volume 2:** 171-172
ZigBee **Volume 1:** 79, 81

OTHER BOOKS IN THE SERIES

Taiwan Cyberwarfare. Information Communication and Electronic Force Command Cyber Warfare Wing. Available on Amazon. US $9.95.

Chinese Radars. Available on Amazon. US $9.95.

Chinese Helicopters. Available on Amazon. US $9.95.

CHINESE SEAPLANES, AMPHIBIOUS AIRCRAFT AND AEROSTATS/AIRSHIPS

- PRODUCT BROCHURES -

WENDELL MINNICK
EDITOR

Chinese Seaplanes, Amphibious Aircraft and Aerostats/Airships. Available on Amazon. US $9.95.

Chinese Rotary/VTOL Unmanned Aerial Vehicles. Available on Amazon. US $9.95.

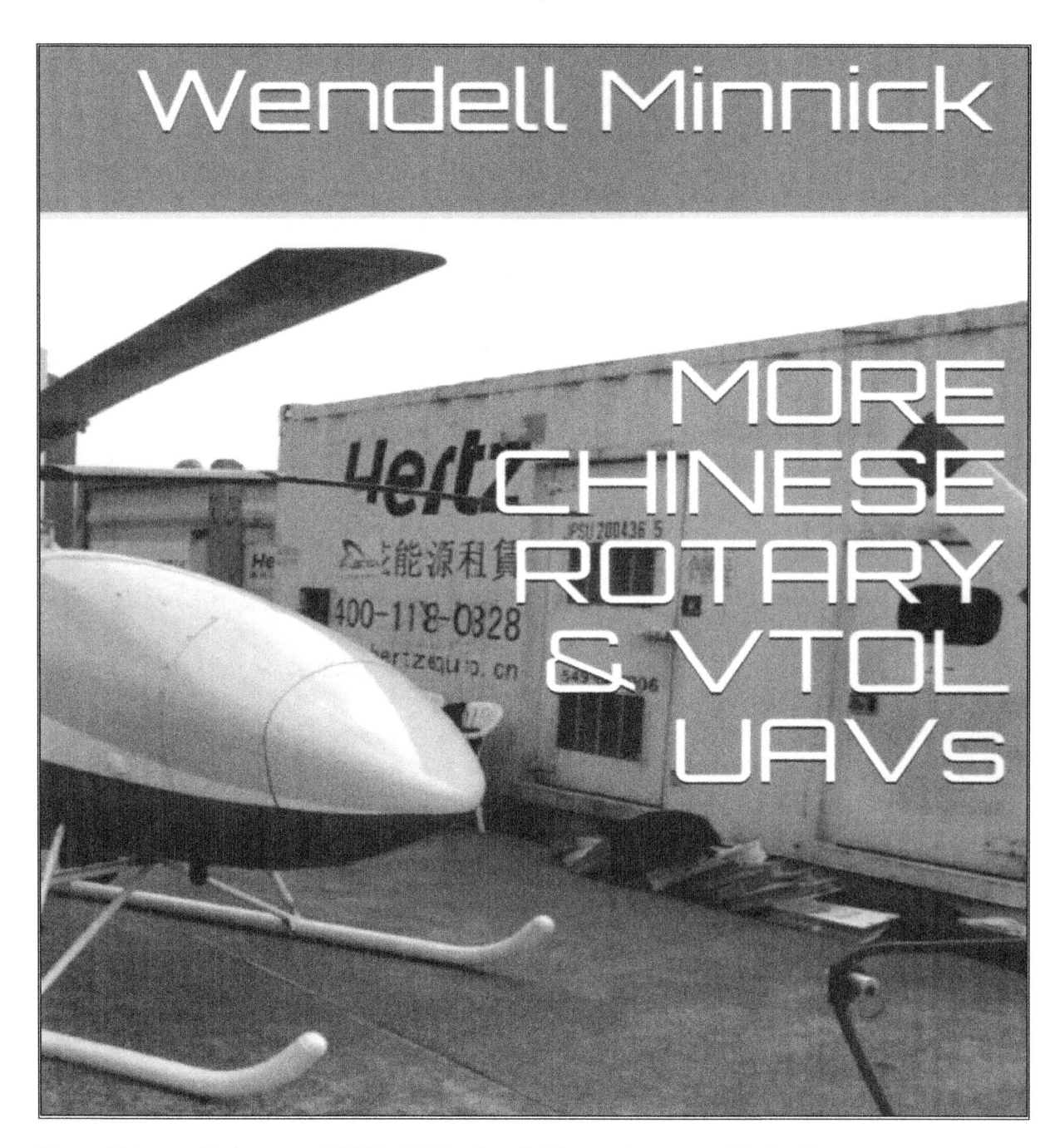

More Chinese Rotary and VTOL UAVs. Available on Amazon. US $9.95.

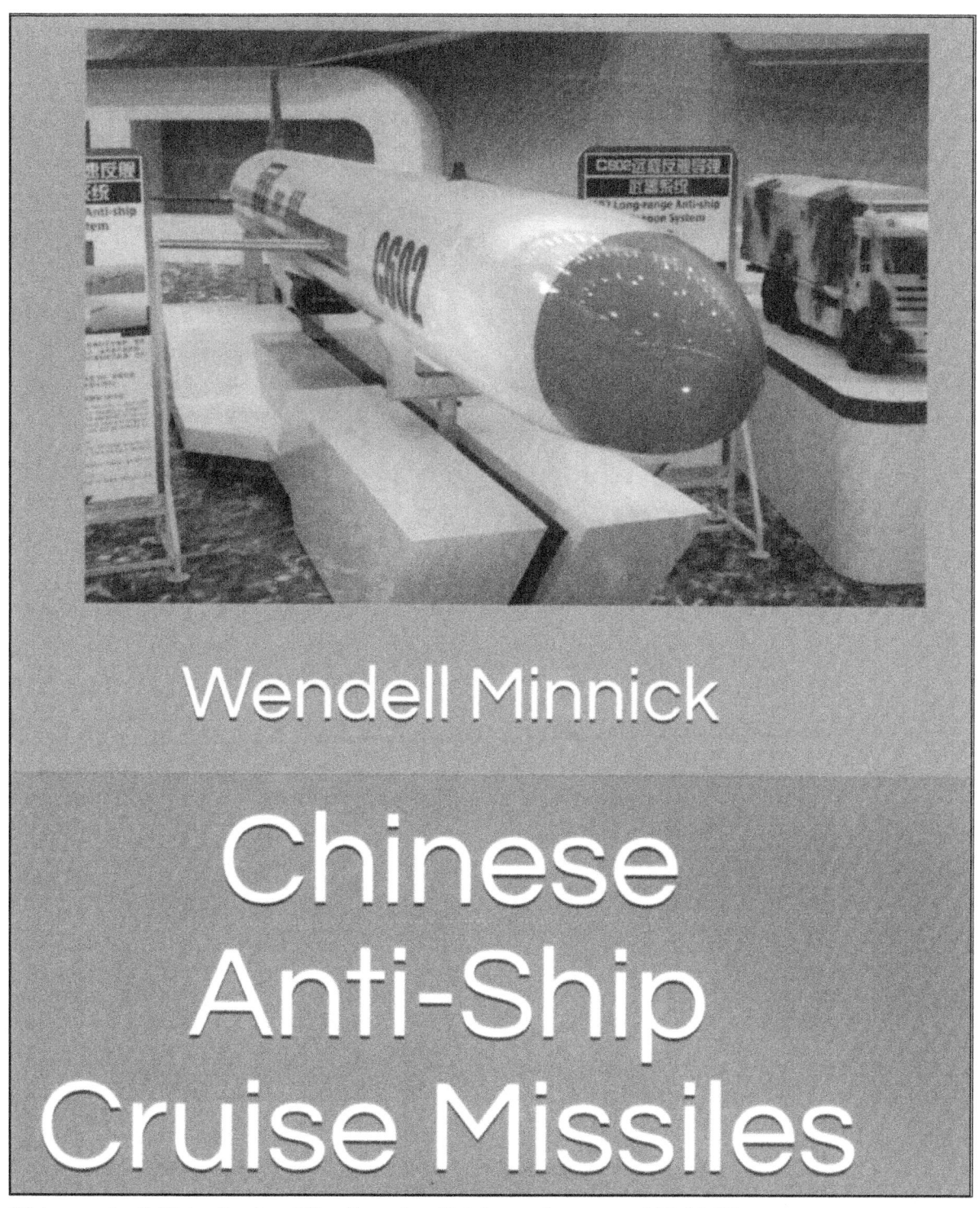

Chinese Anti-Ship Cruise Missiles. Available on Amazon. US $9.95.

Chinese Fixed-Wing Unmanned Aerial Vehicles. Available on Amazon. US $9.95.

More Chinese Fixed-Wing UAVs. Available on Amazon. US $9.95.

Chinese Air-Launched Weapons & Surveillance, Reconnaissance, and Targeting Pods. Available on Amazon. US $9.95.

Chinese Rocket Systems: Multiple Launch Rocket Systems. Available on Amazon. US $9.95.

Chinese Space Vehicles Programs. Available on Amazon. US $9.95.

Taiwan Space Vehicles. Available on Amazon. US $9.95.

Chinese Tanks and Mobile Artillery. Available on Amazon. US $9.95.

List of Foreign Companies and Identities of Taiwan Local Agents. Available on Amazon. US $9.95.

Chinese Submarines and Underwater Warfare Systems. Available on Amazon. US $9.95.

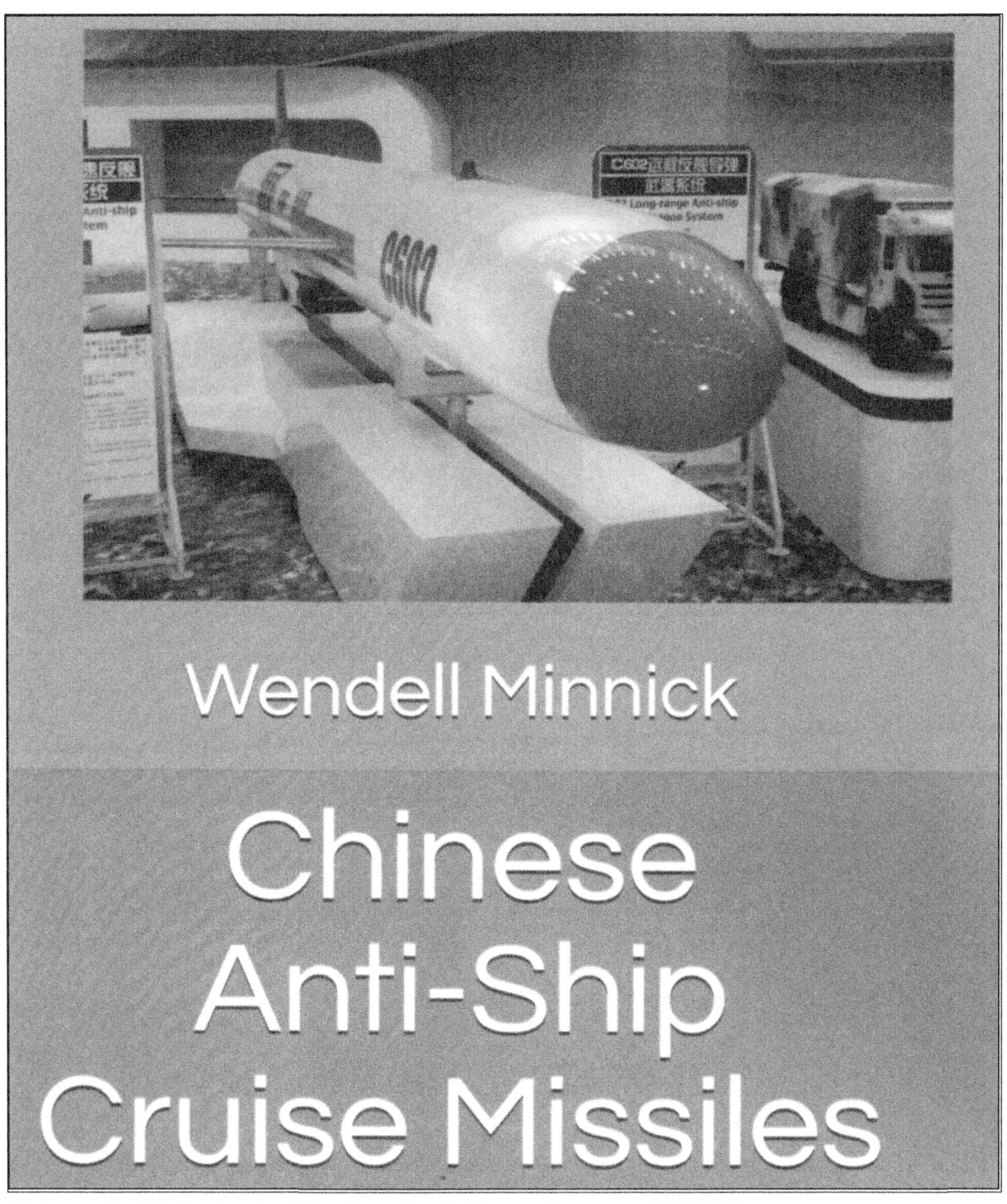

Chinese Anti-Ship Cruise Missiles. Available on Amazon. US $9.95.

CHINESE AIRCRAFT ENGINES

PRODUCT BROCHURES

WENDELL MINNICK
EDITOR

Chinese Aircraft Engines. Available on Amazon. US $9.95.

Chinese Fighter Aircraft. Available on Amazon. US $9.95.

China Market Outlook for Civil Aircraft, 2014-2033. Available on Amazon. US $9.95.

Printed in Great Britain
by Amazon